Biofeedback is a monumental new science which sounds like science fiction. Yet this impeccably researched book attests to the *fact* that a very real and spectacular scientific breakthrough has occurred.

The authors see biofeedback as having an impact on society equal to that of Freud and Copernicus. They envision it as a vehicle for man to find his way back to his origins and become once again master of his own destiny. They detail the actual results of biofeedback training, envisioning its limitless future potential.

Biofeedback may be the next step in the evolution of man. Here is the book that tells you all about it, a book that could dramatically alter your life.

". . . this is certainly the best book of its kind yet published."—*Library Journal*

MARVIN KARLINS was graduated from the University of Minnesota and received a Ph.D. in Psychology from Princeton University. He is Associate Professor of Psychology at City College of City University of New York, and has written a novel and four books in the field of psychology.

LEWIS M. ANDREWS received an A.B. in Psychology at Princeton and an M.A. in Communications from Stanford University. He has written a number of articles in addition to two books in the field of psychology.

BIOFEEDBACK

TURNING ON THE POWER
OF YOUR MIND

MARVIN KARLINS
LEWIS M. ANDREWS

WARNER BOOKS

A Warner Communications Company

WARNER BOOKS EDITION

Copyright © 1972 by Marvin Karlins and Lewis M. Andrews
All rights reserved

Library of Congress Catalog Card Number: 79-39759

ISBN 0-446-92200-5

This Warner Books Edition is published by
arrangement with J. B. Lippincott Company, Inc.

Cover photograph by Jerry West

Warner Books, Inc., 75 Rockefeller Plaza, New York, N.Y. 10019

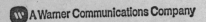 A Warner Communications Company

Printed in the United States of America

Not associated with Warner Press, Inc. of Anderson, Indiana

First Printing: January, 1973

15   14   13   12   11   10

To Mom and Dad
M.K.

To Terri, for contact, both near and afar
L.M.A.

CONTENTS

ACKNOWLEDGMENTS

With a sense of deep satisfaction we acknowledge those who helped make this book possible. Special thanks are due Gay Luce and Erik Peper, who not only gave the authors several hours of interviews but also read and made valuable suggestions on parts of the manuscript. In addition, we are indebted to Gay for her assistance in preparing the Body Rhythm Diary in Chapter 4 and to Erik for making available some of his prepublication research findings. Dr. Elmer and Alyce Green of the Menninger Foundation also deserve special mention; even with their busy schedule they took the time on three separate occasions to cheerfully answer our lengthy inquiries.

Because biofeedback work is so new, some of the most exciting findings have yet to be published. We are particularly grateful, therefore, to the following individuals for granting us personal interviews and sharing their frontier research with us: Dr. John Antrobus, Mr. Jim Beal, Dr. Barbara Brown, Dr. Paxton Cady, Dr. Eleanor Criswell, Dr. Arthur Deikman, Dr. James Fadiman, Dr. Ernest Hilgard, Dr. Julius Korein and his assistant Ms. Lucie Levidow, Dr. Neal Miller, Dr. Joseph Notterman, Dr. Robert Ornstein, Dr. Hal Putoff, Mr. Robert Randle, Dr. Roger Sperry, Dr. Solomon Steiner, Mr. Russell Targ, Dr. Charles Tart, Dr. William Tiller, Dr. Marjorie and Hershel Toomin, Dr. Arthur Vassiliadis, Mr. Marcel Vogel, and Ms. Chris Wendell. We are also indebted to those biofeedback investigators who supplied us with prepublication reports of their work: Dr. Abe Black, Dr. Jasper Brener, Dr. Thomas Budzynski, Dr. Andrew Crider, Dr. Bernard Engel, Dr. Curtis Hardyck, Dr. Thomas Mulholland, Dr. Paul Obrist, Dr. Gertrude Schmeidler, Dr. Gary Schwartz, Dr. David Shapiro and Dr. Johann Stoyva. In the course of our writing we also had the opportunity to interview Adam Crane (a founding director of the BioFeedback Institute), Alexander Everett (president of Mind Dynamics), Dolly Gattozzi (science writer for the National Institute of Mental Health), Lillian Petroni (one of Dr. Elmer Green's first migraine headache sufferers), John Picchiottino (president of Bio-Feedback Systems), and Kenneth Scott (president of Scott Behavioral Electronics). We thank them for their helpful comments. We also received valuable suggestions and/or support from Dr. Larry Bloomberg, Mr. Tony Collins, Dr. Jon Davidson, Ms. Judy Godinger, Mr. Hugh MacDonald, Mr. Stuart Miller, Dr. Michael Murphy, Mr. Allen Rucker, Dr. Harry Schrodet, Mr. Walter Schneller, Ms. Terri Snow, Mr. Jim Tobak and Mr. Chet Wilson.

In doing our research we made extensive use of the university libraries at Princeton, Rutgers and Stanford. We would like to thank the personnel at each of these libraries for their aid—particularly Ms. Eloise Harvey of the Princeton University Psychology Library. Ms. Janet H. Baker did a magnificent job of copy-editing the manuscript. Credit should also go to Vikki Cooper, Anna Jacks and employees of the Princeton Secretarial Service for their fine typing work on early drafts of the manuscript.

Finally, we take great delight in acknowledging our debt to the people at Paul Reynolds, Inc., and J. P. Lippincott Company for their assistance and encouragement during all phases of our undertaking. When one has an agent like Malcolm Reiss he begins to appreciate what elegance in publishing is all about; and when one has an editor like Genevieve Young he begins to understand what elegance in writing is all about. To Malcolm and Gene and Don Bender, Edward Burlingame and Isabelle Holland—thanks for making two authors extremely happy.

12

BIOFEEDBACK

INTRODUCTION

There are many signs that man may be undertaking a systematic exploration of the vast, imperfectly known universe of his own being, a step as epochal as his construction of a science of the galaxies.

WILLIS S. HARMAN, Director
Educational Policy Research Center
U.S. Office of Education at the
Stanford Research Institute

The Voluntarium: Hospital of the Future?

"I want to welcome you all," said the tour guide as he ushered the ten visitors through the front door of the Hans Berger Memorial Voluntarium. "If you'll follow me, I'll try to show you the highlights of our new facility." The guide walked briskly down a long polished corridor, motioning his charges to follow. "If you glance to the left," he noted, hardly breaking his stride, "you'll see a replica of the building's cornerstone. The insignia is particularly interesting. It is Latin for our motto, 'Patient, heal thyself.' "

"Don't they need doctors here?" asked one of the visitors.

"Nothing could be further from the truth." The guide paused before an open elevator and waved everybody inside. "Doctors are as important here as in a traditional hospital. The difference is one of medical philosophy. In the voluntarium doctor and patient work together in a partnership against human illness—a partnership where the patient is given more opportunity to acquire responsibility for, and power over, his own health."

The visitor nodded. "And the term voluntarium. . . ?"

"The term was coined by psychologists in the seventies to describe an institution where patients would learn to combat their medical problems by bringing their bodily functions under their own voluntary control."

The elevator doors slid apart, revealing two diverging floor shuttles. "Take conveyor one, please. . . . I think any further questions you might have will soon be answered," the guide added.

The visitors crowded onto the moving platform. "We're coming up on the biofeedback training ward," commented the guide, pointing to a series of rooms ahead. "This is where patients learn to tune in on their inner states and control them through the power of their own minds." Everyone strained to see as the conveyor passed into the first room.

"What is an anxiety station?" asked a teen-age girl, pointing to a sign on the half-opened door.

"Do you know what anxiety is?" countered the guide.

"Yes."

"Well, this is where people with anxiety learn to overcome it. Look—over there." The guide pointed to a woman resting quietly on a bed attached to some kind of machinery.

"What's she doing?" asked the teen-ager. "And what's that noise?"

The guide smiled. "She's learning to relax by listening to the amplified sound of her breath."

"And that guy over there?" asked the girl's father, motioning toward a patient with a row of electrodes across his forehead.

16

"He's learning to reduce his anxiety, too. But he's doing it by listening to his forehead muscle."

"I see," said the father, skeptically. "And I suppose that his friend in the next cubicle is doing it by listening to the sounds of his hair."

"No." The guide was patient. "He's seeing if he can calm down by regulating his brain waves."

The father grunted disbelievingly.

"Before you come to any conclusion, wait till the tour is over," the guide suggested. "Right now, look into these treatment rooms."

As the voluntarium guests watched, the moving platform carried them past rooms with cryptic labels such as "headache clinic," "hypertension center," "insomnia section" and "heart chamber." In each one patients, wired to machines, sat or lay about listening to electronic squeaks or observing flashing lights.

In one room a woman sat quietly, an intent look on her face. An electrode protruded from the middle of her forehead and the tip of her finger.

"She's trying to raise the temperature in her hand to abort a headache," the guide observed.

In a second room a patient wired to a complex apparatus watched red, yellow and green lights flashing on a panel at the foot of his bed.

"He's trying to regulate his heartbeat," the guide said.

"Tell me," asked an older woman in the group, "do these patients remain on those machines for the rest of their lives?"

"No." The guide was emphatic. "They use them only as long as they need to in learning to regulate their bodily functions. Once they have achieved voluntary control over their internal states they can leave the machine behind."

"Forever?"

"Sometimes they might stop back for a checkup to make sure they haven't forgotten what they learned."

The woman rubbed the back of her neck. "Do they always have to come back here for the tests?"

"Not necessarily. Later on we'll get a look at the outpatient dispensary, where patients are issued portable machines for home use. They can use these—or rent time on machines from local service centers opening around the nation. Of course," the guide hastened to add, "work with some of the machines must be done under a doctor's supervision. Does that answer your question?"

The woman nodded.

"Fine. Now if you'll look to your right you'll see the entrance to the experimental section of the voluntarium. . . ."

In a literal sense this tour is fiction. It has not taken place—as yet. But the treatments described are not imaginary; they have already been used effectively in laboratories around the country, as you will see. The *method* used by voluntarium patients to gain mastery over their bodily functions is not fiction, either. It is currently being employed to help people use the power of their minds not only to overcome medical problems but to achieve fuller, happier, more meaningful lives. The technical name for this method is "biofeedback training," an unwieldy expression, perhaps, but one that is destined to become an essential part of your daily vocabulary.

To our children, biofeedback training will be as commonplace as television has become to us.[1] * In the near future it may be used to "starve" warts and cancerous tumors, regulate glandular secretions, reduce mental illness and increase longevity. Huge corporations like Martin Marietta are already looking into biofeedback as a way to stimulate creativity and reduce anxiety in top-level executives. Biofeedback machines are being developed for industry and education that will tell a foreman or teacher whether an employee or student is paying attention to his work.

On a more individual level, future biofeedback training may be used by men to achieve erections and by women to control ovulation, a rather advantageous combination.

* Superior figures refer to the section of Notes at the end of the text.

And for those contemplating meditation, brain-wave bio-feedback training, where a person learns to control his beta, alpha, theta and delta brain waves, promises a head-start on the road to satori. Durand Kiefer suggests, "Perhaps within another decade the number of alpha-masters, theta-masters and delta-masters living among us may exceed the number of Zen-masters, Yoga-masters, and Sufi-masters who have lived since time began." [2]

Nowhere is excitement over biofeedback training more evident than in the laboratories and universities across the nation. Here, in the privacy of their offices, scientists and scholars are free to give vent to their enthusiasm—enthusiasm they are required to mask in their antiseptically written journal articles. As we listened to these knowledge-able investigators describe their ongoing research and projected programs, it became readily apparent that the only thing more exciting than biofeedback research today will be biofeedback research tomorrow.

The ultimate possibilities for man's self-control are nothing less than the evolution of an entirely new culture where people can change their mental and physical states as easily as switching channels on a television set. [3]

Even the mere prospect of such a culture, let alone its realization, will have dramatic philosophical consequences for our dehumanizing society. For the last century, a mammoth technology has combined with the fatalistic science of psychology to produce an image of man that is both alienating and destructive. Biofeedback research is a direct challenge to this image; it promises to give man the feelings of freedom and dignity that are necessary for a sense of self-worth. One writer has aptly termed the current "systematic exploration of man's inner being" as "no less epochal than man's first step on the moon." [4]

What the Public Doesn't Know about Biofeedback— and Why

Despite the immense potential of biofeedback technology for enhancing man's freedom and self-control, it is

19

difficult for the layman to get an accurate picture of bio-feedback research. Information coming from within the ivory tower is distorted by the academic vices of bickering and one-upmanship. Many scientists spend more time forcing biofeedback training into the terminology of their own pet theories than they invest on original research. The very first meeting of the Bio-Feedback Research Society was split over what to call this new process. The behaviorists wanted to ordain it under the mantle of reinforcement theory. The humanists, on the other hand, wanted to adopt it under the sacred concept of voluntary control.[5] "I was called to this one conference in Washington about biofeedback training," recalls a self-proclaimed humanist proudly. "They originally called their meeting *Operant Conditioning, Something, Something.* . . . I stormed about this so much they ended up calling the conference something quite different to imply voluntary control. It amused me, although I must admit I almost walked out of the thing." On one level, these petty disputes among august academicians are entertaining. Unfortunately, they serve to fragment biofeedback research under different rubrics and into different journals, making it difficult for scientists, let alone laymen, to keep track of advances in the field.

If academic discussions of biofeedback are obscure, claims made by laymen with economic interests in the area are frequently oversimplified, inaccurate or deliberately misleading. "You have a lot of people in the field," explains Robert Ornstein, one of the most creative biofeedback scientists. "Some of them are really doing good work, but you have a lot of people publicizing work they haven't done very well . . . and then there are the people who are really out to make a fast buck."[6] Dr. Thomas Mulholland of the Bedford, Massachusetts, Veterans Administration Hospital is even more outspoken. He believes that recent interest in biofeedback research has created "a huge 'sucker' market for the kinds of gear that are supposed to permit easy recording of bioelectric signals."[7]

Of course, there is also the problem of instant salvation organizations which distort and manipulate journalistic accounts of biofeedback experiments, particularly those focusing on the voluntary control of mental states. These groups offer everything from multiple orgasms for women to professional success for men—at a price, naturally. It is easy to scoff at these modern patent medicine men, but their influence is growing at an alarming rate.

The purpose of this book is to cut through conflicting academic jargon, on the one hand, and cultish exaggerations, on the other. We will attempt to present a clear picture of biofeedback research: its history, its present status and its implications for the future. To report this subject with the care it deserves, the authors undertook the most extensive review of biofeedback research to date. This material was then augmented with scores of personal interviews with pioneering investigators in the field, personal inspections of laboratory facilities, discussions with people who have successfully controlled their internal organs, and personal experience on biofeedback machines.

Some of the material we present will seem fantastic. It *is* fantastic. Nevertheless, every instance of biofeedback training that we mention is clearly documented in the notes to the book. And while you are reading, remember this: some scientists consider *our* position to be conservative.

Understanding
Biofeedback

*There is a new conception of man's capacities form-
ing within the scientific community. During the
twentieth century, academic science has radically
underestimated man—his sensitivity to subtle sources
of energy, his capacity for love, understanding and
transcendence, his self-control.*

DR. JAMES FADIMAN
Stanford University

What Is Biofeedback and Why Is It Important?

Imagine you are visiting your first English pub
and your host challenges you to a game of darts. Never
having played, you graciously decline and then—in the
finest American spirit—run out, buy a set and begin prac-
ticing in your hotel. After the first hundred tosses you
begin getting a feel for the game; by the next day, you're
ready to go out and challenge the Queen's finest.

You have learned your dart game well. But let us pre-
tend you were forced to practice your throws blindfolded,

with plugs in your ears. Could you ever perfect your toss under these conditions? No. Improvement would be impossible because you lacked the vital component of learning: *feedback* concerning your performance. Deprived of visual feedback—unable to gain *knowledge of results* concerning your throwing accuracy—your plight would be hopeless.

We use feedback so regularly in our everyday life that we seldom realize how pervasive and important it is. Yet as one eminent scholar has pointed out, "Every animal is a self-regulating system owing its existence, its stability and most of its behavior to feedback controls." [1] It is only when we are suddenly deprived of our normal opportunity to receive feedback—for instance, in the case of sudden blindness—that we come to understand its momentous value for our very survival.

The term "feedback" is of relatively recent origin, coined by pioneers in radio around the beginning of this century. Mathematician Norbert Wiener, a founding father of research in feedback, concisely defined it as "a method of controlling a system by reinserting into it the results of its past performance." [2] Thus, a dart player learns to control his performance by observing and acting upon the results of his previous tosses.

What then is biofeedback? It is simply a particular *kind* of feedback—feedback from different parts of our body— the brain, the heart, the circulatory system, the different muscle groups and so on. Biofeedback training is the procedure that allows us to tune into our bodily functions and, eventually, to control them. Without such training most of us would never be able to receive feedback from our internal world, feedback that is absolutely necessary if we wish to gain mastery over all aspects of our behavior. Without it we are no better off than the blindfolded dart thrower—unable to observe the results of our probes into inner space. In a typical biofeedback training session a subject is given this feedback by hooking up with equipment that can amplify one or a number of his body signals and translate them into readily observable signals: a flashing light, the movement of a needle, a steady tone, the squiggle of a pen. Once a person can "see" his heartbeats or "hear" his brain waves, he has the information he needs to begin controlling them.

An Example of How Biofeedback Training Works

If it weren't for the metal disks on Eric's throat, he might be mistaken for an ordinary student reading for exams. In Eric's case, however, the test he is studying for is taking place while he reads: he is being examined for "subvocalization," the tendency to silently mouth words while reading. Such a habit limits reading speed to a ceiling of about 150 words per minute while increasing reader fatigue.

The metal disks on Erik's neck are electrodes, small microphones designed to record the minute bioelectric potentials generated by the movement of his vocal muscles. These potentials are fed into an amplifier known as an electromyograph, or EMG machine, and translated into a signal. The signal will tell Dr. Curtis Hardyck, director of the project, whether Eric is subvocalizing. (Before the advent of advanced electromyographic techniques it was literally impossible to detect subvocalization.) When the muscles of Eric's larynx are relaxed, the amplitude of the signal is low. When, however, he talks at a conversational level, the amplitude of the signal rises appreciably. Thus, if the amplitude of the EMG signal is high while Eric is reading silently, we can diagnose him as a subvocalizer.

If Eric's EMG results indicate subvocalization, he will learn to overcome it via biofeedback training. Dr. Hardyck has utilized such an approach with dozens of subvocalizers, and in an overwhelming number of cases the results have been dramatic and swift: patients learn to kick the subvocalization habit in one to three hours! [3]

The method of treatment is relatively simple. The patient is seated in a comfortable chair, electrodes are placed on each side of his Adam's apple and he is given a book to read. He is informed that a tone will come on if certain speech-muscle activity is present. To demonstrate this the patient is asked to whisper. As he does, the tone comes on. He is then told to relax. When he does the tone terminates. The patient is encouraged to turn the tone on and off until satisfied he can control its activation at will. Once he believes he can, the biofeedback session begins. The patient is instructed to read but to keep the tone off as much as possible. In short order the tone remains off and the subvocalization problem is overcome.

It should be emphasized that biofeedback plays the central role in helping readers overcome subvocalization. Dr. Hardyck observes that "attempts to reduce the speech muscle activity by instruction alone were not successful. In the majority of cases the subjects were not aware of their subvocal activity even when told they were subvocalizing, and they were unable to reduce or to control it without the auditory feedback." [4]

Consider the components of Dr. Hardyck's "biofeedback blueprint." A patient has a problem. It is determined that the problem has a clearly defined basis in some specified activity of the body. Through instrumentation, the patient is permitted to monitor that specific behavior—to tune into his inner world. The activity is represented by a signal of some kind, and the patient is instructed to change the signal while he observes it. If the patient is able to do as he is asked, the inappropriate bodily activity will be modified or eliminated. *It is important to note that the patient is never told to "slow down his muscles" or "speed up his heart!"—instead, he is asked to "keep the tone off" or "make the light dimmer."* People are so out of touch with themselves that direct attempts to change a subtle behavior will frequently achieve the opposite of the desired result.

Once you get a feel for Dr. Hardyck's specific treatment procedure, you are well on the way to understanding biofeedback generally. This is because the strategy behind biofeedback training is basically the same regardless of the problem being treated. You will see this quite clearly when, in later chapters, we describe the course of biofeedback training for overcoming a wide variety of health and performance problems.

One needn't use biofeedback monitors to get in touch with the inner self. Yogis and Zen masters have been doing it for centuries. So have serious practitioners of judo, karate and other martial arts. How do they do it? The same way we could if we followed their teachings and developed the same degree of self-discipline, patience and introspective power. Yet how many Americans are willing to invest the time and undergo the rigors of true meditation for a glimpse of the inner light? Very few. For most of us, the path to satori is an organic meal here and there, five minutes in the lotus position before bed, and catching the swami on the late-night talk show.

If we wish to become connoisseurs of our inner world, we must develop the ability to look inside ourselves; to focus our attention inward and listen for signals the untrained ear cannot hear. In the United States this is extremely difficult to accomplish for three reasons. First is the problem of *attitude*. As Westerners we have not been taught to look inward—to feel the blood flow, to hear the heart beat or, as one psychologist put it, "to play upon our internal organs." [5] If anything, we have been led to believe that the area within our skin verges on the disgusting. "Yeecch, who wants to hear their blood flow?" is a comment an American might be expected to make. The bodily processes are something to be tactfully ignored, like an ill-timed belch. Second, there is the problem of *cultural noise*. The signals from our body are exquisitely soft; to hear them one must find a quiet place and devote his entire attention to their detection. It is no accident that serious mediators "contemplate from their own mountain peak." For serious introspection, the silent, uninterrupted life of the hermit is almost a prerequisite. In the United States precious few mountaintops remain. Even if we assume the average American could detect his body signals, how could he be expected to hear them above the din and bustle of his daily life? Nobody has ever heard a pin drop on the Seventh Avenue subway. Finally, there is the problem of *time pressure*. In the United States it is almost a national

27

value to be in a rush. "A quiet moment for meditation" in our country means just that—a moment. The student who once asked the Yogi, "I've got a few moments . . . can you teach me satori?" was serious—and that's the tragedy of it. Learning to recognize the symphonics of our inner self is a time-consuming passion for the serious meditator. That means it is out of reach for most citizens in our clock-conscious culture.

Biofeedback Machines: "Space-Time Demolishers"

Which brings us back to biofeedback training. One can reach inward without relying on instrumentation, but in our society it is very difficult to do so. It would seem more reasonable to strike some form of compromise: use instrumentation (when it is necessary) in the initial stages of biofeedback to get a person into himself; then, once he's pointed in the right direction, let him continue alone.

This is what biofeedback training accomplishes. It works like this. Even unconscious body states give off energy that can be measured. Small increases in hand temperature, for example, will cause more heat to be released from the hand. Similarly, increases in muscle tension will accelerate electrical activity at the surface of the skin. As a stethoscope amplifies heartbeat sound waves, biofeedback machines monitor these subtle energy shifts and translate them into light flashes, clicking noises or some other signal accessible to our normal senses. When several functions are monitored at once, the results can be aesthetically pleasing. Subjects hooked up to special equipment are actually able to produce a light and sound show with their body waves—to "watch the 'music' of their minds and bodies flicker across various screens in a dazzling matrix of colors, each coded to a different function." [6]

Once a person is able to recognize his body waves, he soon learns to control them at will. *Nobody knows why this happens; it just does.* Furthermore, the machines are only a temporary necessity. After a little practice—the time varies, depending upon what body function the person is trying to control—he is able to carry on without mechanical aids. When used in this manner, instrumentation is analogous to the training wheels on a child's bicycle which are discarded once the youngster has learned to keep his balance. If you prefer a more exotic description, think of a biofeedback machine as a "time-space demolisher," [7] allowing us to achieve control over our nervous system in a shorter period of time than would be possible without it. A yogi can learn to control his brain waves in a matter of years; the average person using biofeedback training can learn to control his in a matter of hours.

It also seems that instrumentation is eminently compatible with the American character. Americans are more at home when they are using machines. Many people who would never be interested in controlling their inner states under normal circumstances suddenly sit up and take notice when they find out that a gadget is involved. Some investigators have been quick to recognize the value of instrumentation in getting Americans interested in exploring "inner space." Dr. Eleanor Criswell, director of the Humanistic Psychology Institute, put it this way: "It has been said that Americans need gadgets in order to be able to do things, so if this is America's gadget way of giving itself permission to meditate, then it's worth it." [8] Finally, it should be remembered that instrumentation is valuable because it sometimes offers the only means by which a person can monitor certain kinds of nervous system activity (for example, brain waves).

The Greening of the Nervous System

Every once in a while a discovery is made that changes the course of scientific thinking and makes us revise our conception of "what is." Copernicus made such a discovery; so did Darwin and Freud. What of biofeedback? Is it a discovery of sufficient magnitude to change the face of science and our conception of reality? It is still too early to know for sure; yet the initial impact of biofeedback

29

research is undeniable: it has already forced us to revise our conception of the entire nervous system.

At New York's Rockefeller University, Professor Neal E. Miller sits at his well-papered desk oblivious to the boat traffic that passes by his window on its way up the East River. He is an intense man—and his intensity is directed in the pursuit of scientific knowledge. Nobody appreciates more than Dr. Miller just how far biofeedback research has forced us to revise our view of the nervous system. He knows because, as the master architect of that revision, he fought tooth and nail to have his unorthodox views accepted. When he began to do battle, his ideas were considered so incredible that, in his own words, "for more than a decade it was extremely hard for me to get any students, or even paid assistants, to work seriously on the problem. I almost always ended up by letting them work on something they did not think was so preposterous." [9]

At the time Dr. Miller began his revolutionary work, it was generally believed that man was regulated by two distinct nervous systems: the voluntary and the involuntary. The voluntary, or somatic, nervous system includes the nerve cells and fibers that serve the skeletal muscles. It is responsible for all arm, leg and jaw movement, for changing posture—in short, for all movement that we normally regard as deliberate or "consciously controlled." The involuntary, or autonomic, nervous system involves the eye pupils, heart, blood vessels, stomach, endocrine glands and all functions traditionally considered automatic or "beyond our control."

Miller has accomplished no less than the termination of this distinction. He has done this by showing that the so-called involuntary system can, indeed, be brought under voluntary control. Working with rats paralyzed by curare (to eliminate responses from the "voluntary" nervous system) Miller was able to show conclusively not only that animals could learn to control their "automatic" responses but that they could do so with a great degree of specificity.

One rat, for instance, learned to blush in one ear at a time! [10]

Recently, it has been found that man, too, can learn to exert conscious control over such autonomic functions as heart rate and blood pressure. Evidence of this revolutionary new human capacity doesn't surprise Dr. Miller unduly. "I believe that in this respect men are as smart as rats," he says, smiling.

The Return of the Mind

Biofeedback work has created a second, equally important revolution in scientific thought: the return of the mind. Scientists, particularly psychologists, hesitate to study things they cannot observe and measure. For many decades this hesitance resulted in a virtual rejection of human consciousness as a proper domain for scientific inquiry. Only today are we coming to understand the tragic price of this embargo in terms of its cost to scientific knowledge and human dignity.[11] Fortunately, in the wake of biofeedback research findings, this embargo is being lifted and the investigation of man's consciousness is becoming a legitimate scientific enterprise.

Biofeedback training has opened a doorway into our minds. It has given us a technology and a procedure for exploring our innermost thoughts and feelings while allowing scientists to peek over our shoulders and record what we experience. The man who has learned to control his internal states becomes, in a way, an organic scientific instrument—a human barometer for measuring the internal atmosphere. He becomes, in a deeply satisfying sense of the word, a tool for carving out a science of human consciousness.

Biofeedback Training: What's in It for You?

Not long ago one of the authors attended a biofeedback training session where several persons were learning to

31

increase their output of alpha brain waves, a state of consciousness some people find relaxing and pleasant. After the session a few of the participants were informally asked to describe what for them was the major value of biofeedback training. "It made me feel good," was the first respondent's reply. "A chance to experience something new, something I might not be able to experience any other way," suggested a second person. "Without the aid of drugs," added a third participant. "It's kind of a personal ecology trip—you don't despoil your internal environment with artificial ingredients; you use only what you already have."

The achievement of a pleasurable state. An opportunity to explore untapped frontiers. The chance to practice personal ecology. All important advantages of biofeedback training. But perhaps the greatest advantage is one we don't normally think about. It was described most aptly by a fourth participant, who said simply, "It gave me a sense of dignity."

It is difficult to capture on paper the sense of exhilaration felt by the person who, through biofeedback training, is able to master his body through his own self-power. All his life he has been taught he cannot control his "involuntary" nervous system. All his life he has been encouraged to "let the pill do the job" or "go ask Mr. Jones for the answer." All his life his capacity to control his own body, his own mind, his own fate has been challenged, deprecated, even ridiculed. And then, suddenly, there he is controlling his own nervous system—a man on his own magical mastery tour. "My God, I really did it by myself, didn't I?" one patient proudly announced after she learned to control her body waves. "Yes," answered the doctor, "and that's the beauty of it."

Biofeedback Training
for Your Health

> . . . *biofeedback promises to return us to a more holistic kind of medicine in which the patient will acquire more responsibility for, and power over, his own health, no longer finding himself treated as a defective organ, but as a person in a context, with a life style and habits that affect his own body. Biofeedback puts the emphasis back on training, rather that the "miracle pill" or surgery, and indicates that the mind itself can be trained to do most of the things that mind-changing drugs are used for.*

GAY LUCE and ERIK PEPER

The Placebo and the Shofar

We've all seen the plot on TV. The elderly Mrs. Smith is dying and knows it. She summons her doctor and asks him how long she has to live. The physician hedges but finally estimates "a week . . . maybe two . . . a month at the outside." The woman gets terribly upset, saying she is ready to die but not before her fiftieth wedding anniversary, a little more than two months away. The doctor says he will do everything he can to help her remain alive that long; then he goes downstairs and tells Mr. Smith his wife will be dead in a matter of days. Nine weeks later we see Mr. and Mrs. Smith celebrating their anniversary. The next day Mrs. Smith is dead.

The story of Mrs. Smith is tearjerking fiction at its best, an old woman heroically staving off the grim reaper until she can cherish a treasured moment. The question is: How fictional is the story of Mrs. Smith?

To the Jewish people Yom Kippur, the Day of Atonement, is the most solemn religious holiday of the year. It is a day of great importance, the day they "heed the call of the Shofar" and ask God to forgive them for their sins and transgressions. It is a day that Jews anxiously await, much as Mrs. Smith awaited her anniversary day. Sociologist David Phillips discovered something extremely interesting about Yom Kippur and the people who observe it. Studying the mortality records for Jews in New York and Budapest, he found a notable drop in the death rate just before the Day of Atonement. There was no such drop among non-Jews before the High Holy Day. Carrying his investigation further, Phillips also examined the mortality pattern around people's birthdays. What he discovered tied in nicely with his Yom Kippur findings: there was a significant dip in deaths before birthdays and a significant peak in deaths thereafter—which all means, according to Phillips, that "some people look forward to witnessing certain important occasions and are able to put off dying in order to do so." [1] It seems that Mrs. Smith is not the only person who keeps death waiting. [2]

The Jews in Phillips's study were able to observe Yom Kippur by postponing their deaths until the holiday was over. Of course, this "will to live" has its limitations: death is inevitable and there is only a finite amount of time by which we can postpone it. Yet the idea that man may be able to put off death *at all* raises some interesting questions, questions like "How does man do it?" and "Might he have other powers that can affect his general state of health?"

34

For many years doctors and scientists have suspected that man might have more control over his nervous system than was readily admitted in textbooks of physiology and psychology. There is, for example, the disturbing problem of the placebo. A well-known example of a placebo is the "sugarcoated pill," a capsule the patient believes contains medicine when, in fact, it does not. Oftentimes, the patient swallowing these bogus tablets recovers from his illness as well as the patient who receives real medicine. Why? There are many answers suggested: the patient's faith in the doctor dispensing the pills; his belief in the effectiveness of his "medicine"; the "psychological" basis of his condition. One fact remains, however, regardless of the answer: without the aid of "real" medicine the patient has overcome his illness. In other words, the patient has cured himself.[3]

One of the major contributions of biofeedback research has been to demonstrate that some of the placebo effects we have attributed to a patient's faith in the doctor and some of the death postponements we imputed to willpower might be grounded in factors far less mystical. The patients might simply be exercising a certain amount of voluntary control over their health.

In the coming years we can expect to learn even more about man's capacity to control his own nervous system. Psychologist Gardner Murphy predicts that "within the next decade or two certainly a very significant control of cardiovascular and gastrointestinal responses may be anticipated, not only with immediate clinical values in bringing in or shutting out various classes of bodily information, but with the deeper scientific value of giving a much wider view of what the human potentialities for such inner experience and such inner control may be."[4]

The Patient as a Prescription: Medical Applications of Biofeedback Training

Once we realize that man has the capacity to play an active role in combating his own medical problems, we

35

must give him every available assistance in making that role as effective as possible. This is where biofeedback training comes in: it is the best tool we have to help man achieve control over those nervous-system activities related to his physical well-being. Using biofeedback training, doctor and patient can cooperatively mount an assault against human illness, an assault where the patient becomes his own prescription for good health. Some medical problems will, of course, still have to be treated by traditional procedures (drugs, surgery, etc.), but for those amenable to treatment by biofeedback procedures the patient will be afforded the opportunity to acquire more responsibility for, and power over, his own health.

As the reader familiarizes himself with the medical applications of biofeedback training, he should keep in mind that all the training was conducted under medical supervision. Even the most well-meaning individual can do himself harm if he tries to regulate his internal states without proper instruction and guidance. This should surprise no one. The person attempting to regulate his nervous system without proper supervision is not unlike the individual who takes drugs without a doctor's prescription: in both cases the person is trying to manipulate his internal states without the knowledge necessary to do so safely.[5]

In the following section we will review the ways in which biofeedback training is now being used to combat different medical problems. If, while reading, you get a sense of *déjà vu*, don't be overly surprised. You met some of these facts once before—on a "fictional" tour of a voluntarium.

Combating Anxiety

For most people anxiety is one of two kinds. There is the anxiety that is "free-floating," a general feeling of apprehension that is not attached to any specific type of activity. It is best characterized by the person who says, "I've felt uneasy all day . . . and I don't know why." A second kind of anxiety is more specific: it occurs while a person is undertaking (or anticipating) certain types of behavior he dreads, like taking an exam or looking down from high places. This kind of anxiety is reflected in the comments of a man waiting for an elevator: "Every time I get into one of these damn things I feel closed in. . . . I get a sense of foreboding—like I'm not able to breathe." Anyone who suffers from claustrophobia understands what it means to be anxious.

Most doctors agree that one way to reduce both free-floating and specific anxiety is to get patients to relax. The problem is—how? Instructing the individual to relax sometimes backfires: persons asked to relax particular muscles sometimes respond by unintentionally tensing them instead. A procedure is needed where the patient can receive feedback about his relaxation efforts.

Respiration biofeedback, in which subjects listen to their amplified breath sounds, is a promising new procedure that reportedly helps patients relax more effectively than they could following verbal instructions.[6] Recently, Dr. Paul Grim studied the impact of respiration biofeedback on anxiety in ninety-five nursing students. Each student was first asked to take a psychological test which measured her general level of anxiety. Then she was ushered into a small but comfortable room where the actual biofeedback session was conducted. During the session she lay quietly on a bed, hooked up to the equipment that amplified her breathing, tried to relax her body by letting go of all her muscle tension, and simply listened to magnified sounds of her own breathing. When the session ended, each student was asked to take the psychological test a second time.

What was the effect of respiration feedback on the nurses' behavior? First, their breathing seemed to become smoother and more regular. More significantly, when Dr. Grim compared the women's test scores from before and after the biofeedback session he found that their anxiety had been significantly reduced.[7] He also discovered that the students enjoyed their experience; a few even asked

37

him if they could repeat the session after their exams were over.

Dr. Grim, recognizing the importance of relaxation for the reduction of anxiety, suggests that respiration biofeedback might be advantageously used in therapy. Dr. Thomas Budzynski and his colleague Johann Stoyva at the University of Colorado have taken him up partially on his suggestion—but they use muscle biofeedback rather than respiration biofeedback as the therapeutic tool.

In a typical Budzynski training session, electrodes are attached to a patient's forehead so that any movement in the frontalis (forehead) muscle can be detected, amplified and fed back to the patient through an EMG (electromyograph) machine. The task of the patient is to keep the frontalis muscle relaxed. He does this by controlling a tone which rises when the muscle contracts and falls when it relaxes.

Dr. Budzynski has chosen the frontalis muscle for biofeedback work because it is a crucial barometer of the patient's relaxation level. Once a person can relax the frontalis "he will usually relax his scalp, neck and upper body. . . . [The frontalis] usually contracts when people are agitated or tense, although most people are not aware of it." [8]

The Budzynski biofeedback training procedure is valuable in the treatment of generalized anxiety because it helps patients learn to relax. It can also be used in overcoming specific forms of anxiety. In one remarkable case Dr. Budzynski reports that he "liberated a 22-year-old girl from an unbelievable list of phobias, including panic attacks, fear of heights, fear of crowds, fear of riding in cars, and claustrophobia. When systematically confronted with the feared images in a relaxed state, she gradually lost her intense feelings about them. It is almost impossible to remain highly anxious when deeply relaxed." [9] In two recently reported case studies, Budzynski and Stoyva used EMG biofeedback successfully to help a forty-two-year-old management consultant combat his public speaking anxiety (which had forced him to refuse several lucrative speaking engagements) and assist a forty-five-year-old

38

woman in overcoming her extreme anxiety at social gatherings.[10]

Some investigators have suggested that brain-wave biofeedback might also be effective in combating anxiety. They point out that since some persons report a relaxed, pleasant experience while in the alpha brain-wave state, being able to switch into the alpha state at will might be useful in helping the highly anxious individual calm himself. Nobody is suggesting that people in an alpha state would be without anxiety—instead, brain-wave control is seen as a way individuals could keep their anxiety within manageable bounds. Dr. Martin Orme, director of experimental psychiatry at the University of Pennsylvania Medical School, suggests that

> alpha might be used to bring down the level of a person's anxiety to a point where he can function at his best. We all need a certain amount of anxiety to function. It is well accepted that we function best as anxiety rises to a certain point on a bell-shaped curve, and past that point we do increasingly worse as anxiety increases. If alpha can be used to knock down anxiety to the point on the curve where we work most effectively, it can be a most important development.[11]

Does Dr. Orme think that this development will ever come to pass? He cautions us that his speculation "is three levels or more from where we are now, but it is something to consider."

At this point in the developing area of biofeedback research, no apprehensive executive should expect to lick his anxiety with a fifteen-minute alpha session before the next board meeting. Yet there are already reports of individuals using brain-wave control to come down after a hard day's work. The ultimate value of brain-wave biofeedback in reducing anxiety must still be determined. Meanwhile, biofeedback training in respiration and particularly in muscle relaxation present promising procedures for combating debilitating states of anxiety.

39

"I had spots in front of my eyes, terrible nausea and a pounding headache over one eye. I would try to stay up and do things, but I simply couldn't stay on my feet. I couldn't tolerate light . . . I couldn't tolerate noise . . . I could not talk to people. I was just in constant, agonizing pain and I would have to go to bed." So spoke Lillian Petroni, recalling her migraine headaches.[12]

As time passed Mrs. Petroni's migraines got worse. Drugs provided temporary relief but they were no panacea; there was the troublesome problem of side effects. "I took medicine but it made me dopey," she recalls. "It impaired my skills and limited the kinds of things I could do. And," she added, "a drug is just a pain-killer—it didn't help cure my headaches." [13]

In 1968 Lillian Petroni heard about the work of Dr. Elmer Green at the Menninger Foundation in Topeka, Kansas. (She was a research assistant at Menninger at the time.) "My husband Frank told me about it," she remembers. "He was talking with Dr. Green one day and said I was home suffering from a migraine. Green said he might be able to help." [14] Shortly afterward, she went to see Dr. Green, hoping for a new weapon to battle her migraine adversary. One can appreciate her hope: Mrs. Petroni had been fighting a losing battle for five years!

The hope was not unwarranted. Under the tutelage of Dr. Green, Mrs. Petroni learned a biofeedback procedure that helped her overcome her migraines. How long did she need to use biofeedback equipment to learn this voluntary control? Two weeks! At latest report Mrs. Petroni has been free of migraine headaches for three years.

When one listens to Mrs. Petroni discuss her biofeedback training in Dr. Green's laboratory, it is obvious that the experience has significantly altered her life. "Besides making me healthier, it has made me a lot more understanding of myself and the needs of others," she maintains. It is not an idle claim. Since leaving Kansas and moving to Tucson, Arizona, Lillian Petroni has undertaken an ambitious project to help others through some of the techniques she learned at Menninger.[15]

Just what kind of biofeedback training did Lillian Petroni receive that was so effective in treating her problem? The same type that was recently given to twenty-eight additional headache sufferers under the direction of Dr. Green and his associates, Dr. Joseph Sargent and E. Dale Walters. Here is how the program worked. Each participant was first given a complete physical examination. This was followed by the actual biofeedback training procedure. Temperature-sensitive electrodes were placed on the midforehead and the right index finger of each participant and attached to a "temperature trainer" meter which registered forehead and hand temperature differences. The patient's task was to move the meter indicator, by "mental power," to the right, signifying he had increased the warmth of his hand in comparison with his forehead. Each patient was given a list of phrases to help him relax and accomplish his task ("I feel quite quiet" . . . "I am beginning to feel quite relaxed" . . . "My feet feel heavy and relaxed" . . . "My arms and hands are heavy and warm" . . . "My whole body is relaxed and my hands are warm, relaxed and warm" and so on).

Why should the ability to keep the hand warm and the head cool help a migraine sufferer? The answer is not definite yet, but it seems that "migraine . . . headaches might be ameliorated by regulation of blood flow"[16] between the head and the hand.

At first, patients required the meter feedback as a guide to accomplish their handwarming response. After several sessions, however, they were able to elevate the temperature *without* the meter altogether. Once they could do this they were in a far better position to abort their migraines. Now, each time they felt a headache coming on, they could activate their handwarming response—even when no temperature trainer was available.

Although the temperature trainer treatment is still in the experimental stage, it does seem to hold promise for migraine sufferers. Not all the participants in the program enjoyed Lillian Petroni's spectacular success, yet 80 percent of the migraine sufferers were evaluated as improved after biofeedback training.[17]

Not every headache sufferer is afflicted with migraines. There are also the so-called tension headaches which result from chronic muscle contraction (usually brought on by emotional conflicts). Dr. Green and his colleagues tested some of these headache sufferers with his temperature biofeedback procedure, found it not particularly effective and suggested that Dr. Budzynski's muscle-relaxation procedure might be more valuable.

We have already discussed Dr. Budzynski's biofeedback training procedure in the treatment of anxiety. As will be recalled, the task of the patient, using EMG biofeedback, was to keep his forehead muscle relaxed. The same method is employed in the treatment of tension headaches. This is because relaxation of the frontalis is usually accompanied by relaxation of the neck and scalp muscles which, when tense, produce the headaches in the first place.

Using muscle-relaxation biofeedback training Dr. Budzynski helped five patients (a research technician, two middle-aged housewives, a young high-school teacher and a middle-aged businessman) reduce their tension headaches.[18] Training, to be effective, required four to eight weeks of patient participation—both at home and in the laboratory. The time seemed well worth it: besides learning to control their headaches without the need for drugs, the patients also learned to recognize the onset of tension and how to reduce it in their daily lives, even when a biofeedback machine wasn't present. The treatment was so beneficial that a "former grammar-school teacher, who had been forced to give up teaching because the kids gave her headaches, was able to resume her profession." [19]

What about the tens of thousands of Americans who suffer from headaches and look to men like Dr. Green and Dr. Budzynski for aid? Can they be helped like Lillian Petroni or the schoolteacher who was able to return to the academic wars? Further research will be required before such questions can be conclusively answered. Yet the initial findings do look highly encouraging, and we suspect that biofeedback training in headache control will be a major medical treatment in the coming years.[20]

Combating Hypertension (High Blood Pressure)

If you walked into Dr. David Shapiro's laboratory at the Harvard Medical School and looked around, you might think you were witnessing a patient's physical examination. After all, the young man seated nearby was obviously having his blood pressure checked; the inflatable cuff was already wrapped snugly around his upper arm. If, however, you lingered in the room awhile you'd soon realize that this examination was like none other you'd ever seen. Who ever heard of a patient having his blood pressure tested twenty-five times in one sitting—and this to the accompaniment of flashing red lights, strange tones and even an occasional slide of a *Playboy* nude? Better yet, who ever heard of a patient's blood pressure going *down* under such conditions? Yet that is exactly what happened when Dr. Shapiro and his colleagues asked a group of young men to lower their blood pressure through biofeedback training.[21]

The experiment worked this way. A subject was ushered into a light- and sound-controlled cubicle, seated, and a cuff for measuring blood pressure attached to his arm. A crystal microphone was placed inside the cuff to amplify blood pressure activity from the subject's brachial artery. Once he was comfortable the patient was told to watch for a blue light signal. This would signify the beginning of a test trial where he was expected to keep on a flashing red light and tone as long as possible. The more frequently the light flashed, the better the subject was at controlling his blood pressure. Every time the subject accumulated twenty light flashes he was rewarded with a five-second peek at a *Playboy* pinup. Under such experimental conditions, subjects rapidly learn to lower their blood pressure.

43

The Shapiro findings have been replicated in other scientific laboratories. At the University of Tennessee, for example, Jasper Brener and his colleagues find that subjects can readily lower their blood pressure by using their "mental processes." In an improvement on the Shapiro work, these investigators measure blood pressure from the finger rather than the upper arm, because the arm cuff "was found to produce considerable discomfort to subjects when used for prolonged periods." [22]

Why are men like Shapiro and Brener so concerned about teaching subjects to lower their blood pressure through biofeedback training? The two men are in exact agreement when it comes to the answer: to provide the patient with a possibly effective method for treating hypertension. It is estimated that approximately one out of every eight Americans suffers from hypertension (elevated blood pressure without a demonstrable cause), a condition that doctors implicate in strokes and heart attacks. Because prolonged periods of high blood pressure can damage the blood vessels and arteries, it is important to find ways to lower the pressure. Drugs have been used for this purpose in the past—but dosage levels and side effects can be a problem. It is the hope of men like Shapiro and Brener that biofeedback training will augment or replace drug therapy in the treatment of high blood pressure in the future.

Initial results look promising. In an experiment involving actual patients, Aimee Christy and John Vitale of the San Francisco Veterans Administration Hospital report that blood-pressure biofeedback training was successful with two labile hypertensive patients.[23] At Harvard, Dr. Herbert Benson has helped five of his seven patients lower their blood pressure in the laboratory with biofeedback. "Nevertheless," as Dr. Benson notes, "to generalize this training—and keep blood pressure down at home and at the office—may require some revision of behavior and even re-evaluation of life style, as well as some professional behavior therapy." [24]

"To sleep; perchance to dream," said William Shakespeare. But to the bleary-eyed insomniac the statement might better read, "To dream; perchance to sleep." Insomnia is a particularly frustrating affliction, trapping the hapless victim in restless, sleepless night prisons. Sometimes drugs can help the insomniac achieve the sleep he craves, but the price is high. Some of the drugs are habit-forming, others have side effects. For instance, recent evidence has suggested that certain sleeping compounds (the "hypnotic drugs") suppress certain parts of the sleep cycle that might be important to psychological well-being.[25]

We already know that people have learned to relax by controlling their muscles. There is also evidence suggesting that some individuals can learn to achieve a pleasant state of calm by controlling their brain waves. Might there be a way to combine body-wave and brain-wave feedback training to achieve the sleep state?

Yes, there is, and once again Dr. Budzynski has done it. The biofeedback training program was divided into two parts. Insomniacs first learned to relax by frontalis-muscle feedback.[26] Once this was accomplished they were hooked up to an electroencephalograph and taught, via brain-wave biofeedback, to produce those wave patterns associated with the onset of sleep. "With this training people who had taken four hours to fall asleep were dropping into slumber twice in a 20-minute lab session." [27]

There is currently a growing interest in biofeedback training as a means of overcoming insomnia. In California Dr. Maurice B. Sterman has evidence suggesting that cats can produce longer epochs of undisturbed sleep by learning to control brain waves from the sensorimotor cortex area of their brain.[28] Might this apply to humans as well? We don't know yet. There are reports that many individuals use brief alpha brain-wave sessions before retiring as a way to get to sleep faster. And, of course, there is the continuing work of Dr. Budzynski.

Where will all this work lead us? Dr. Budzynski gives

this buoyant response: "I think doctors are going to lend patients an EMG machine instead of handing out sleeping pills in the future." [29]

Combating Muscular Tics

Patients with spasmodic tics have been helped by EMG biofeedback. Erik Peper reports one case of a young man beset with severe facial tics who tried EMG biofeedback after psychiatric aid had failed. The training procedure utilized a portable EMG machine which allowed the patient to "hear" his forehead muscle contract by clicks delivered through earphones. The patient's goal was to reduce the number of clicks, thereby relaxing the frontalis and, it was hoped, reducing the number of tics.

Looking back on his first session, the patient reported, "It started with my constant concentration on my back. I kept saying to myself relax back, relax back. Then I concentrated on the clicks. At that moment a whole new world was brought before me. I felt a deep warmth. I felt a very heavy feeling of sinking into the bed. I, for once, controlled my body movements, they did not control me. . . . I cried for happiness because I never felt I could conquer my bodily movements. I told myself, 'Don't tic'— and I didn't. I felt a completely calm feeling. The warmth of my body overwhelmed me." [30]

The power of EMG biofeedback training in helping patients overcome tics and other neuromuscular disorders is just beginning to be appreciated. As research continues we can expect new treatment improvements and applications. One such application was recently reported by Dr. Harold Booker and his associates at the University of Wisconsin. It involved helping a middle-aged woman rehabilitate facial muscles damaged in an automobile accident. The task was difficult. The patient had lost voluntary control of the muscles on the left side of her face and was unable to close or blink her left eye. Further, conventional therapy had been unsuccessful in helping the woman relearn use of the affected muscles. To make matters worse,

46

the woman was very upset because the muscular problem had distorted her facial appearance and given her an abnormal look.

Using a graded biofeedback training program (where the woman first learned to regulate one muscle and then several at a time), the patient was gradually able to regain voluntary control of her facial muscles and achieve a nearly normal facial appearance. The total program—involving practice at home and in the laboratory—spanned several months, but the woman didn't seem to mind. "[She] was quite interested and enjoyed the program, so that her motivation to carry out the treatment was high," comments Dr. Booker. As for the value of EMG biofeedback, Dr. Booker is as enthusiastic as his patient. He concludes his report optimistically, suggesting that the method "can be more generally applied to similar clinical problems" in the future.[31]

Combating Heart Disease

Dr. Jasper Brener and his associates at the University of Tennessee; Dr. Paul Obrist and his colleagues at the University of North Carolina Medical School; the Shapiro group at Harvard and the Engel team in Baltimore; Dr. Peter Lang at the University of Wisconsin; Dr. Neal Miller at Rockefeller University; Dr. Leo DiCara at the University of Michigan—the list could go on and on, for the area of human cardiac control seems to be the most fascinating of all when it comes to voluntary regulation of the nervous system. There is something about man learning to "drive his own heart" [32] that seems to bespeak the ultimate adventure and, in this sense, makes a patient who can control his heart more intriguing than a doctor who can transplant one.

Literally scores of scientists have established that man can, through force of mental discipline alone, regulate his own heartbeat.[33] At this writing, however, only two published studies have actually used heart-rate biofeedback training to help patients with heart trouble. Both were conducted by Dr. Bernard T. Engel and his colleagues, Dr. Theodore Weiss and Dr. K. Melmon, at the Baltimore City Hospital. The findings of these studies are promising and point to the day when biofeedback training might be used as part of a treatment regimen to aid patients in preventing and combating certain kinds of heart disease.

In their major study Dr. Weiss and Dr. Engel were interested in seeing whether eight heart patients could learn to control dangerous irregularities in their heartbeat by force of mental discipline alone. The dangerous irregularities are called "PVCs"—premature ventricular contractions—and it is important to control them, for as Weiss and Engel note, "the presence of PVCs is associated with an increased probability of sudden death." [34]

Heart-rate training took place in a quiet, windowless laboratory. The patient, lying comfortably on a hospital bed, was hooked up to a cardiotachometer which converted his heartbeat into electrical signals. These, in turn, were fed into a computer, analyzed and translated into red, yellow and green lights on a panel at the foot of his bed. Watching his light panel "traffic signal," the patient was told to "drive" his own heart by following the "rules of the road": slow his heartbeat when the red light was on and increase it when the green light appeared. The patient's goal was to keep his heartbeat at a safe "middle speed," signaled by a steady yellow light.

Using his biofeedback traffic signal the patient first learned to speed his heart, then slow it and, finally, to keep it beating within narrow normal limits. "Some of Dr. Engel's patients have achieved a 20 percent speeding or slowing of their hearts—about sixteen beats a minute from an eighty-beat-per-minute base. This self-willed rate change in one direction or the other tends to even out the irregular beats. Why? Researchers are not quite sure, but it works." [35]

Perhaps the most exciting part of the Weiss and Engel work is what happens when the patient is disconnected from his training equipment and sent home to fend for himself. After extensive training on the machines he finds

48

he can take his lessons home with him and regulate his heartbeat without the need for artificial biofeedback. One of the eight patients has sustained her low PVC rate for almost two years, reporting that she is able to detect and modify her irregular heartbeat while at home. Three other patients have also been able to use their biofeedback training to regulate their PVC activity away from the laboratory. Most patients can't describe just how they accomplish their cardiac control, although one man said he imagined pushing his heart to the left and a woman pictured herself swinging back and forth. Dr. Engel explains the control this way: "It's like an athlete who does something well. He's grooving." [36]

Learning to regulate cardiac arrhythmias through the power of the mind is not an easy task. Many hours must be spent in getting a feel for PVCs and how to control them. Drs. Weiss and Engel are optimistic about reducing training time in the future. "It should be possible to accomplish comparable results in much shorter periods of time," they claim, "as more nearly optimal conditioning technics are employed." [37] Even now, however, the many hours spent in training seem worth it. As Gay Luce and Erik Peper emphasize, "training may be fundamentally tedious, and may require a long time, but unlike drugs it gives the patient a sense of mastery over his own body. One of the women in Dr. Engel's group was elated when she could report that she no longer had dizzy spells or thumping in her chest because of her own ability to cope with the symptoms." [38]

Raising Havoc with Our Own Nervous System

By showing us that man has the power to control his own nervous system, biofeedback researchers have also alerted us to the danger that such power may be inadvertently or intentionally used in ways not conducive to good health. It might well be that many psychosomatic illnesses arise through man's own manipulation of his cardiovascular and gastrointestinal responses. Doctors have long suspected that man's "state of mind" might have something to do with a wide variety of maladies such as ulcers, asthma, hypertension and tuberculosis. Their speculation is well founded. Biofeedback research tells us that a person can indeed use the power of his mind to regulate his body.

The problem, once we realize this, is to harness man's internal control in the service of good health, making sure, at the same time, that the individual doesn't unintentionally or unwisely employ his power to raise havoc within his own nervous system. Take the case of Mrs. X. Suffer though she did, Mrs. X needed her migraine headaches.[39] She hated to entertain, but her husband liked to bring friends home for dinner and during the weekends. Afraid to confront him with her feelings, she unconsciously developed convenient migraines as an excuse for getting rid of guests and, ultimately, for not having any company at all.

Almost as soon as she entered the Menninger Foundation's program for treating migraines through temperature control, Dr. Sargent and Dr. Green suspected that something was wrong. Each time Mrs. X started to meet with some success in controlling her headaches, she mysteriously began to fail again. "Her unconscious realized that she was threatening her way of being," recalls Dr. Green. It was a vicious circle. The pain motivated her toward successful control of the headaches, but the anxiety of losing her excuse drove her to failure. Finally, the strength of biofeedback training overcame her defenses and allowed the fear to surface to consciousness. "I know what my problem is," she confided to Dr. Sargent one day. "I hate to have company." [40]

Instances such as this one should not be surprising. The very existence of biofeedback training shows that the line between physical and mental is a hazy one. If "mental power" can change physical states, it is not unreasonable to assume that many, if not most, physical ailments have psychological antecedents. The case of Mrs. X does serve, however, to suggest some possible dramatic uses of biofeedback training in psychotherapy. In the next few pages we will discuss a few of the imaginative ways scientists have already incorporated biofeedback into the therapeutic situation.

The Enemy Within

You have certainly heard the expression, "Man is his own worst enemy." Dr. George B. Whatmore, a private physician in Seattle, Washington, not only believes this statement but has spent the last thirty years showing how man can negotiate a psychological peace within himself using biofeedback training.[41]

How do people turn against themselves in the first place? To explain this process, Dr. Whatmore has coined the descriptive term "dysponesis"—"*dys* meaning bad, faulty, or wrong, and *ponos* meaning effort, work or energy." Essentially, Whatmore argues that people defeat themselves by unknowingly misdirecting their energy. For example, a young actor called upon to perform for the first time will often brace himself by increasing his heart rate, blowing up his blood pressure, and flooding his system with adrenalin and other potent hormones. Not only does this reaction sometimes interfere with effective acting and create discomfort, but over a long period of time it can lead to pathological tissue deformity. Other common examples of dysponesis include such obvious psychosomatic problems as exaggerated fears, ulcers, impotence and frigidity.

A person with dysponesis is not stupid; he is simply fallible. This disorder comes about in two ways. First, a person might make a self-defeating response yet still be successful. The actor who was inappropriately bracing his nervous system might still have played his role well. Then, because he unconsciously associated good acting with his misdirected efforts, he might continue to repeat the error at every performance, perhaps to the point of paralysis. If you doubt that people make such mistakes, consider the prevalence of superstition, even in modern times. Joe rubs a "lucky" rabbit's foot before applying for an important job. He gets the job. Did the rabbit's foot convince the employer that Joe was the right person for the vacant position? Probably not, but Joe will still rub the moldy hare's paw before every important decision.

Dysponesis also occurs when a person continues to expend energy in a response that was once appropriate but no longer serves any purpose. Dr. Whatmore gives the example of umlauted vowels in the German language. When these learned muscle sequences are carried over into another language such as English they become a hindrance; and that person will have a German accent.

Dr. Whatmore claims that dysponesis is resistant to standard forms of psychotherapy. "Patients we have studied who have undergone prolonged psychoanalysis and other forms of psychotherapy have shown no significant reduction in their dysponetic patterns. . . . Hypnosis cannot correct dysponesis because it cannot teach a person new skills. . . . Drugs can sometimes diminish symptoms resulting from dysponesis . . . but they do not alter the underlying response tendencies of the organism and often lose their effect in time." [42] Only biofeedback training, Dr. Whatmore asserts, can provide an efficient method for eliminating dysponesis.

How does Whatmore's procedure work? Let us briefly review the case of Mrs. C, a fifty-two-year-old housewife who suffered from severe anxiety and depression. Nine times during the last twenty-five years she swung down into acute depression, the last six requiring hospitalization. When she arrived for biofeedback training, she had lost over twenty pounds in the preceding two months. "She cried much of the time and felt there was no hope for her," recalls Dr. Whatmore. "She considered herself a perfectionist and ordinarily had much energy, but now she dreaded each day."

Treatment (Dr. Whatmore terms it "effort training") began with thirty-seven sessions directed at identifying and controlling her self-defeating bracing efforts. Several monitoring devices similar to the EMG were attached to different parts of her body—legs, abdomen, shoulder, forearm. Each time she made an unnecessary bracing effort to solve a problem or difficulty, the excess energy she expended was picked up and registered on a meter, earphone or some other readout mechanism. Gradually, she learned to bring these "depressing" responses under voluntary control. It is not clear whether Mrs. C's bracing produced the fatigue which led to her depressions, or vice versa. However, by attacking the "response tendency" produced by Mrs. C's illness, the illness itself was ameliorated.

The final stage of training involved teaching Mrs. C to recognize and control faulty behavior in her daily life. She learned, for example, not to overreact to situations such as "failing to achieve one of her perfectionistic goals." Finally, after 188 sessions she was able to lead a normal, pleasant life. As an objective measure of his technique, Dr. Whatmore recalls that "whereas [Mrs. C] had been hospitalized six times . . . prior to effort training, no hospitalizations have been necessary in the nine years since effort training began." [43]

For most of his professional life, Dr. Whatmore has not received the credit he deserves for his innovative work. Fortunately, like many other pioneering scientists, he persevered in his efforts. Only today is the value of his techniques becoming fully appreciated. We have already pointed out how Dr. Budzynski and his colleague, Dr. Stoyova, are now using EMG biofeedback to achieve a wide range of beneficial effects in patients.[44] They are currently considering ways in which muscle feedback might facilitate other kinds of anxiety-reducing therapies. For example, by attaching EMG electrodes to his client's chest, a psychiatrist could objectively determine if the therapy sessions were arousing too much anxiety or too little.

Of Censorship and Muscle Cramps [45]

Aside from helping alleviate the more severe clinical problems of anxiety and depression, biofeedback training also promises some relief for normally neurotic people. Everyone, for example, suffers in varying degrees from the problem of self-deception. By this we mean that people are not as objective as they would like to believe. Each person systematically screens out certain kinds of useful information, either because it violates social norms or upsets his self-image. In war, for example, a soldier will not allow himself to see the civilian targets of rifle bullets as "innocent women and children." To preserve his self-image, he must view these unfortunate victims as "the enemy," undifferentiated from the military force he opposes. A more common instance of self-deception is the case of the plump young lady who avoids looking at herself in mirrors.

As psychologist Gardner Murphy pointed out in his address to the first Annual Meeting of the Bio-Feedback Research Society, self-deception is usually accomplished by body movements that prevent unwanted information from reaching the conscious mind. These body movements are similar to the muscular defenses that Wilhelm Reich once termed "character armour." In Murphy's words "the striped musculature of the arms, hands, trunk, neck, and by implication, other parts of the body, may be conceived to be used all the time in the battle of thought, especially the battle *against* thought, specifically the battle against recognition of information, and most of all, against information unfavorable to the self." [46] Irregular breathing is a very common, though unconscious, way to avoid clear contact with unwanted information and is a frequent symptom of those people who complain of feeling cut off from the outside world. By slowing respiration and thereby reducing oxygen intake, they effectively dim awareness. Many psychiatrists have noticed, for example, that when a client begins to deal with a repressed conflict, his breathing suddenly becomes shallow. This is quickly followed by a complaint of feeling tired out or incapable of coping. Knowledgeable therapists will simply say, "Breathe!" Often the problem will immediately begin to emerge into consciousness.

How can biofeedback training help people overcome self-deception? Murphy suggests that people could easily learn to monitor and overcome their unconscious censorship with EMG-like devices. Electrodes could be placed over muscles commonly associated with information blockage. Areas that are abnormally tense or that tighten up too often would give off sufficient bioelectrical feedback to permit voluntary control of the muscles. "Going through such a discipline," Murphy predicts, "it should be possible for any reasonably cooperative human being to learn, by the monitoring meter technique, where he is getting tense, and to learn how to relax, and to learn to take advantage of new information which comes to him." [47]

Not only might biofeedback eliminate self-deception but it could also be used for reducing deception *between* people and simultaneously strengthen mutual trust and sensitivity. Researcher Tom Mulholland imagines the following scene in a future encounter group where the participants wear flashing pins that monitor heartbeat. A man is introduced to a woman, and immediately his "heart pin" begins to pulsate for everyone else to see. Mulholland speculates that the woman would be embarrassed at first, but then her light would also begin to blink at a faster rate. Seeing this, the man would begin driving his heart even faster until the couple resembled a movie marquee. "It would be of great interest to discover if such procedures could teach people to have a much greater sensitivity to . . . behavior which reflects emotional states," Mulholland says.[48]

Rx: The "Executive BF-1000"

Another probable use of biofeedback training in therapy will be the prevention of stress-related disorders, such as ulcers. Just as medical clinics now prescribe preventive medicine regimens, our voluntarium of the future would encourage all executives over thirty-five (and anyone else in high-pressure positions) to take a standard stress-reduction training program. The format for this helpful innovation will probably parallel already existing EMG techniques for allowing problem solving and relaxation to occur simultaneously.[49]

A typical session might go as follows: Mr. G, a harried executive, enters a pleasantly decorated room in a quiet wing of the voluntarium. A friendly "facilitator," most likely a person with a paramedical education, helps Mr. G off with his clothes and then attaches variously colored electrodes to different parts of the patient's body. Slowly, Mr. G sinks back into a plushly padded, motor-controlled water bed. As the warm waves lull him into progressively deeper levels of relaxation, he tries to clear his mind as best he can. After fifteen minutes, he is instructed to start thinking about problems of the day. Whenever the EMG meter tells the facilitator that tension levels are too high, Mr. G is instructed to stop thinking about problems and to concentrate on lowering the biofeedback tone. Once he has returned to a low level of agitation, the facilitator asks him to resume problem solving. This simple process is repeated until Mr. G can handle everyday crises without throwing his nervous system into an alert status.

Medical Biofeedback Training: A Look down the Road

Medical biofeedback research is in its infancy, yet it has already changed the face of medicine and the body of knowledge concerning the mind of man. We are coming to understand, however faintly, man's awesome power to change his own internal destiny. We are catching a faint glimmer of insight into the possibilities that lie before us in the realm of mind over matter—of mental discipline over physical disease. How far will biofeedback research carry us down the road to the elimination of human illness? No one is sure, but most agree that the distance will be impressive.

Some of the most exciting new research is being conducted by scientists working on the frontiers of brain-wave biofeedback. In Boston Dr. Thomas Mulholland is studying biofeedback as a possible tool for diagnosing brain disorders.[50] Dr. Joe Kamiya of the Langley Porter Neuropsychiatric Institute in California elaborates on this idea: "Someday it might be possible to examine a patient's physiological states and diagnose his neurosis just as the physician now detects tuberculosis by examining an X-ray." Furthermore, he speculates, if healthy mental states can be defined, "people can be trained to reproduce them. Instead of gulping a tranquilizer, one might merely reproduce the state of tranquility that he learned by . . . [biofeedback] training." [51]

In Beaumont, Texas, at the Angie Nall School for problem children, voluntary control is already being used as a substitute for tranquilizers in hyperactive children. Initial results indicate that alpha brain-wave training helps these youngsters calm down and control such problems as stuttering and insomnia. One child, dubbed "the runner" because he ran all the time, was able to slow down to a fast walk once he learned alpha control.[52] Alpha brain-wave training is also being used by Erik Peper to help persons overcome the experience of pain (one girl reportedly used it to good advantage at her dentist's office).[53]

In several laboratories biofeedback monitoring is being used to train epileptic patients to suppress paroxysmal spikes, an abnormal brain wave. Dr. Julius Korein, a leading investigator in this area, reports one case where a woman used brain-wave control to quash an oncoming epileptic seizure. Yet he is quick to caution that this is an extremely unusual case and at the present time control of epilepsy through biofeedback training is still an aspiration, not a reality.[54]

Finally, a growing number of scientists are speculating that brain-wave biofeedback training might help some individuals produce mental states so satisfying they can serve as a substitute for drug addiction. As an alternative to the drug experience, the "brain-wave high" might seem to pale by comparison; yet some people do report very gratifying experiences in, for instance, the alpha state. For these persons, perhaps, the drugless brain-wave trip might be a feasible possibility.

Brain-wave biofeedback is not the only area being explored on the roadway to human health. Scientists working on the frontiers of electromyographic (EMG) biofeedback have also made some exciting incursions into our understanding of nervous-system control by mental power. Living by the motto "If it moves—monitor it," EMG researchers have attempted biofeedback training with almost every muscle in the human body. The future of this research looks promising; just recently, for example, EMG biofeedback was used to facilitate muscle relaxation in persons with neck injuries,[55] to treat a girl suffering from hysterical deafness,[56] to help stroke victims regain use of their paralyzed muscles [57] and even to help performing opera singers! [58]

Whether it be brain-wave, body-wave or some other type of biofeedback training, researchers are coming to appreciate the far-reaching implications of such a procedure for man's health. Some scientists envision the day when ulcers will be controlled or eliminated by training patients to limit the secretion of their gastric juices.[59] Others suggest that "voluntary starvation and absorption of cancerous growths through blood flow control might be found to be feasible." [60] Such voluntary regulation of blood flow might also permit man to combat maladies such as Raynaud's disease, a medical problem where the patient has poor blood circulation to the hands. Dr. Green knows of two cases where people have successfully turned their hands from blue to red by using biofeedback training.[61]

How far will biofeedback research carry us down the road to the elimination of human illness? The gap between where we are today and where we might be tomorrow—the distance between today's experimental programs and tomorrow's fully operational voluntariums—is vast indeed. Yet the road is before us, and the journey has begun.

The Inner Trip

The most significant thing that may be facilitated through training in the voluntary control of internal states is the establishment of a Tranquility Base, not in outer space but in inner space, on, or within, the lunar being of man.

ELMER GREEN, ALYCE GREEN
and E. DALE WALTERS

A Whole Inner World Is Awaiting Discovery

Until recently, man could catch only fleeting glimpses of the unlimited universe that is his mind. William James, the eminent turn-of-the-century psychologist, spent a major portion of his life studying altered states of consciousness, yet lamented that "my own constitution shuts me out from their enjoyment almost entirely." [1] Like so many explorers of the mind, James was relegated to the role of secondhand observer. For more than fifty years since James, research into altered states of consciousness has consisted of little more than scattered reports of mystical experiences, reminiscences of spaced-out drug users, and a lot of speculation.

Biofeedback training has created the first revolution in the study of human consciousness. No longer is consciousness a hapless victim of chance or mood or chemistry. Through biofeedback training man is learning to *choose* his state of being. He can explore new experiences in a systematic, controlled way without relying on unstable and often dangerous drugs. He can change his mind without losing his head.

"So what?" you ask. "Who wants freaky hallucinations anyway?" This is certainly a logical question within the context of our culture. The phrase "altered states of consciousness" has come to be associated with hippies, dropouts, popular cults and other questionable groups. Altered states are also associated with words like "meditative" and "mystical," which have always had a ring of unreality to Western ears.

In fact, altered states of consciousness are very much a part of each person's daily life. Sometimes the changes from one state into another are relatively slow, as when we slip from waking consciousness into sleep. Other changes can be more drastic. For example, we can be driving down a freeway in a drowsy, daydreaming reverie when the sound of a passing truck will suddenly jolt us into an alert consciousness. Our conscious perception of time is always changing. We sit in the office at work and the minutes may seem like hours, but if we are having a good time at a party the hours may dissolve into minutes. Altered awareness simply refers to the wide range of experiences and perceptions that each of us is capable of generating. In this chapter we hope to show not only that biofeedback training can help control different states of consciousness but that this fluency has a very practical value.

For most of us, life is a series of anxieties. Will I make this sale? How can I lose more weight? Will he (she) like me? Even if we can stop troubling ourselves, our friends and neighbors obsessively maintain a high level of chatter, noise and pollution. We hunger for a respite in our crowded, hectic world, but there are few wilderness retreats in the glass and steel jungle we call civilization. As quiet space becomes increasingly scarce in the external world, we are forced to seek solace in the quiet inner spaces of our minds. Through biofeedback training, we can discover these places and learn to draw strength from them.

The Frequency of Things to Come

Although biofeedback training of consciousness is a recent development, credit for pioneering work under adverse circumstances should go to Hans Berger, a German scientist who discovered the existence of brain waves in the period after World War I. Berger had always believed that the brain emitted some kind of electrical energy and that this energy was associated with specific states of consciousness. Although his contemporaries were actively hostile to this new idea—not a rare situation in the history of science—Berger spent over two decades trying to prove his theories. The breakthrough came in July of 1924. After sticking two electrodes from a crude galvanometer onto the scalp of a seventeen-year-old mental patient, Berger noticed that the meter indicator registered an electrical response. After five more years of research, Berger was able to establish conclusively the existence of two distinct brain-wave patterns, which he named "alpha" and "beta." He also showed a relationship between brain-wave patterns and mental states: beta seemed to be associated with concentration, such as doing an arithmetic problem, while alpha always seemed to accompany states of nonconcentration, or passivity. Unfortunately, the Nazis did not appreciate either the value of Berger's work or his independent spirit. In 1938, he was rudely dismissed from his post at the University of Jena, an incident which led him to suicide three years later.

Since Berger's first successful experiment, the technology for studying brain waves has improved enormously. Researchers now use what is commonly called an EEG machine or, more technically, an electroencephalograph. In simplest terms, the EEG machine consists of three parts: a set of electrodes, which are attached to the scalp with a harmless paste, a brain-wave amplifier and, finally, a device for recording or displaying changes in brain-wave pattern. This recording device is usually a row of inked pens pressing against a continuously unwinding roll of graph paper. The pens oscillate in tune with the brain's changing electrical rhythm, tracing out a brain-wave picture which can be studied, analyzed or hung on the wall as pop art.

Using this advanced machinery, science has expanded Berger's original description of brain-wave activity from two to four patterns: alpha and beta, plus "theta" and "delta." Beta is the highest brain wave, meaning that it has a frequency greater than fourteen cycles per second, and is usually associated with normal waking experience, such as reading this book. Just below beta is alpha (eight to thirteen cycles per second), a state most often described as pleasant, passive and relaxed. Next down the psychic scale is theta (four to seven cycles per second), an intriguing rhythm since it is associated with both creative hallucinations and, occasionally, anxiety. Delta (one half to six cycles) occurs almost exclusively during sleep.[2]

The exact meaning and interrelation of brain waves are still not entirely clear. At present, scientists think that brain waves are the product of electrochemical activity in the brain cells.[3] When these cells, which have the function of combining and transmitting information throughout the body, begin firing in a synchronous pattern, they present a rhythmic activity at the surface of the scalp. When this activity is between eight and thirteen cycles per second, we call it alpha; four to seven, theta; and so on. The brain is rarely dominated by a specific frequency. At one moment, the right hemisphere of the brain may be generating alpha while the left is in beta. A minute later, the frequency readings may be reversed.

Technically speaking, beta is not a smooth rhythm but a flurry of electrical static. The presence of beta usually means that we are using our brain to get something done, like making up a shopping list, studying for an exam, balancing a checkbook or trying to impress a date. In contrast, alpha appears to be a slowing down of electrical discord into a pulsating hum which sweeps regularly over the brain cortex, usually from front to back. (The alpha rhythm is recorded most prominently from metal electrodes pasted to the back of the head.) This turning down of the brain would explain why alpha is generally associated with feelings of calm passivity and distortions of time and space. According to this theory, theta and delta would represent even slower rates of cortical synchronization. This fits nicely with the experience of theta, which is frequently described as a drowsy Kafkaesque state, and of delta, which is sleep. Physically, the trip from beta to delta is a rhythmic unwinding; psychologically it is experienced as a quieting of the mind.

A Minor Revolution

Like so many important innovations, the discovery that man could control his brain states came almost by accident, as a by-product of an investigation with a different purpose. In 1958, Dr. Joe Kamiya was conducting sleep research at the University of Chicago when he decided to try an unusual experiment. In his own words, "I became fascinated by the alpha rhythms that came and went in the waking EEG's and wondered if, through laboratory experiments with this easily traced rhythm, a subject could be taught awareness of an internal state." [4]

To find out, Kamiya wired a subject to a remote EEG machine and placed him in a darkened room. The subject was informed that a bell would ring at different intervals, sometimes when he was in alpha, other times when he was not in alpha. His task was to guess which state (alpha or not alpha) he was actually in. After each guess, he was told the correct answer. "The first day, he was right only about 50 percent of the time," Kamiya reported, "no better than chance. The second day he was right 65 percent of the time; the third day, 85 percent. By the fourth day, he guessed right on every trial—400 times in a row." [5]

Kamiya repeated his experiment on eleven different subjects. His results, aside from confirming his initial experiment, provided a surprise. Not only could the human guinea pigs learn to identify the alpha and non-alpha states, but *"they were able to control their minds to the extent of entering and sustaining either state upon our command"* [6] (italics ours).

Needless to say, Kamiya's work precipitated a minor revolution in experimental psychology. To identify your brain-wave states: that was interesting. But to control them! That was downright fascinating. The problem now was finding an easier and less time-consuming method for teaching subjects brain-wave control. Kamiya himself found the answer: biofeedback. He devised circuitry that would translate the occurrence of alpha waves, as measured by the EEG machine, into a tone. Each time a subject produced alpha waves, he would simultaneously generate a tone. When he stopped, the tone would also stop. The subject had only one task: find some way to keep the tone on as long as possible. As it turns out, most people trained in this manner can learn to produce alpha at will in a relatively short period of time (usually a few hours). It also seems that those who learn brain-wave control quickly are more than good subjects—they also seem to be more emphatic and enjoyable to be around. "I generally tend to have more positive liking for the individual who subsequently turns out to learn alpha control more readily," reports Dr. Kamiya. "I find this especially true of females, for some reason!" he adds.[7]

In the decade since Kamiya's pioneering experiment, the research on training brain waves has accelerated geometrically. Not only alpha, but beta, theta and even delta have come under voluntary control, at least in the laboratory. The machinery has also been modified. The tone can be amplified to indicate increases in brain-wave intensity, or it can be replaced by another form of feedback, such as a light. The basic technique, however, remains the same: the brain-wave impulses, which elude our normal consciousness, are piped through EEG machine electrodes, amplified by delicate circuitry and finally translated into light, sound or some other medium that is accessible to the senses. Once tuned into himself, almost anyone can learn to identify specific brain-wave states and, in short order, to control them.

But wait a minute. If specific brain waves are associated with specific states of consciousness, why do we need feedback? If we are in the alpha state we should be able to experience it directly. Why do we need this elaborate biofeedback setup to tell us what we are feeling?

There are two answers to this question. First, a surprising number of people really do not know *what* they are feeling. Our culture does not teach us to develop the sensitivity needed to detect subtle shifts in our emotional chemistry. To feel means to be aware, but most Westerners "feel" by jolting their nervous systems beyond the conditioned state of civilized numbness with television, drugs, alcohol and loud music. These devices stimulate, but in the long run they reduce the ability to feel.

Second, brain-wave experiences are not the same in all individuals. While brain waves can be measured objectively by the number of cycles the brain generates per second, the *subjective* experience of particular frequencies varies according to the individual. "If you have alpha it may mean a good state or it may just mean eight to twelve cycles per second," says Sonoma State College psychologist Eleanor Criswell.[8] One person's experience of alpha can be another person's experience of theta. Although alpha is usually correlated with feelings of relaxation and well-being, this is not always the case. For some people, the production of alpha can even be an upsetting experience. They have spent so much of their time keeping busy that slowing down from beta into alpha is a source of anxiety and apprehension. The effects of biofeedback training apply to the statistical majority; they are not universal. When we speak of the benefits of alpha brain-wave training we are making a generalization based on what usually is the case; there are always exceptions to the rule.

Since the experience of alpha is not a constant, people require biofeedback training simply to learn what alpha means for them. Once in touch with alpha, people variously describe it as "alert," "nonvisual," "spaced out," "passive," "anxious," "frightening," "ecstatic," "quiet," "high," "letting go," "submissive" and "pleasant." Some find alpha so exciting, they try as hard as they can to become full-time alpha subjects. New York University researcher Erik Peper received a phone call at 2 A.M. from an excited would-be volunteer. Peper was perturbed but not surprised. "Wherever EEG feedback is studied," he claims, "unpaid, uninvited volunteers go to great lengths for the opportunity to tune into alpha." Peper also reports sighting students walking around New York City replete with headset and earphones, monitoring their own brain waves.[9]

A swami, on the other hand, tried alpha and reported to the surprise of onlooking scientists, "I've got news for you. This is nothing." [10] There is even evidence that some alpha experiences are strictly a function of the subject's expectations. One person who came for alpha training already knew that alpha was supposed to be associated with special experiences. As he willingly set out to produce alpha in the laboratory, he reported all the appropriate subjective phenomena—"I'm losing track of space and time"—as well as "There's a rabbit in here so real I can almost touch it." A suspicious scientist turned off the feedback tone while the young man was *still* in alpha. Thinking that he was no longer in the alpha state, the subject stopped having his extraordinary experiences.[11]

Unfortunately, the popularity of brain-wave research has nurtured a cult of the alpha high, which regards the alpha experience as an intrinsic good. Doubtless there are times when the alpha state is, indeed, an advantageous state to be in. Yet, generating alpha, even if pleasant, is not always a good thing to be doing. It is very nice to be relaxed and blissful when you are lounging around your house, but it is not very functional when you are in a situation that demands concentration, such as driving a car, crossing a street in heavy traffic, conducting an important business transaction or figuring out your income taxes.

How does biofeedback training give people the power to control their own brain waves? Nobody really knows. We can only observe that the process is distinctly different from anything else we know about. Consider alpha again. Ninety percent of the people reading this book can produce alpha waves simply by closing their eyes. Other common techniques include opening your eyes in the dark while simultaneously straining to look at something, opening your eyes in the daylight and staring at an object, carefully focusing on a moving target, and simply becoming sleepy. Simple eye-muscle tremors will also cause alpha bursts. Some people have their own peculiar ways of producing alpha. One student does it by holding his breath; another shows alpha bursts as he scores touchdowns on the football field. Dr. Marjorie Toomin, a clinical psychologist, recalls one patient who could only produce alpha while talking. "As soon as he shut up, there was nothing there." [12]

Biofeedback *training* of alpha is something quite different from any of the foregoing. It seems to cultivate an ability best described by Elmer Green as "passive volition." After a certain amount of feedback, people are able to generate alpha simply by willing it, without any effort in the normal sense of physical exertion. In fact, one of the least effective ways to generate alpha is by trying to do so. Typically subjects will walk into a biofeedback laboratory geared up for the anticipated alpha breakthrough. They contort their faces, conjure up wild sexual fantasies or try physical exercise. After what can be hours of fruitless effort—one young man even tried making out with his girl friend while wired up to the EEG machine—the subject gives up and, at that instant, it comes. It is the old Zen paradox. One successful alpha producer reports a typical experience: "I try to forget I'm trying and let myself drift, and the moment I formulate what is happening, in order to remember, to tell it to the experimenter—that moment my experience fragments and the tone ceases." [13]

There are two major practical uses of brain-wave bio-feedback. First, scientists can use biofeedback to construct objective physiological measurements of specific moods and experiences. Mental states such as hate, ecstasy, depression and even affection may be monitored and plotted on a graph. For years poets have asked the question, "What is love?" Soon we may know the answer in exact cycles per second.

Again, the relationship between experience and brain-wave states is not perfect, but it is close enough to suggest some interesting possibilities. Classrooms could be equipped with individual EEG monitors so that professors and teachers could tell when their students were paying attention to their lessons. Hollywood already uses feedback devices on volunteer audiences to test the effectiveness of pilot TV programs and commercials. Noting that serious international crises may be due to negotiators having different thought patterns, Dr. W. Grey Walter, a leading physiological researcher, suggests that "perhaps a diplomat should have his alpha-type [alpha brain-wave pattern] endorsed on his passport." [14] Donald Gould, writing in the *New Statesman,* goes one step further: "[Alpha-type variation] is only one discovered physical difference in the machinery of understanding which can corrupt empathy and communication. There are certainly others. We should study the intimate workings of the brains of the men who discuss our destinies, and match like with like, or at least provide interpreters, and to hell with the shape of the table, or who sits where." [15]

The second potential use of brain-wave biofeedback is control. Once we know how a specific experience translates into EEG feedback, we can attempt to turn the process around, teaching people to reproduce a desired state of consciousness by self-generating the appropriate brain-wave patterns. The extent to which an individual will be able to guide his consciousness at will is still under investigation. We have already indicated that most people can produce states of tranquility through gross alpha training. In the following pages we will discuss research on volitional control of other mental states, showing in detail how man might use biofeedback training to explore the infinite consciousness of his finite mind.

Rational vs. Aesthetic Consciousness

If you take a human brain and surgically cut the *corpus callosum* that connects the left and right hemispheres, an interesting thing happens: the two sections begin to operate independently. This was amply demonstrated in a bizarre experiment where a scientist first severed the two hemispheres of a monkey's brain, then conditioned each side to perform contradictory tasks. One section had learned to press a bar in order to get a banana reward; the other had learned to expect a shock after touching the bar. When both hemispheres were in the presence of a bar, a grotesque fight ensued between the right paw (connected to the left side of the brain) and the left paw (connected to the right side.) One part of the monkey's brain tried to press the bar; the other struggled desperately to stop it.[16] A similar phenomenon is exhibited in humans who have undergone split-brain surgery in order to curb uncontrollable seizures. Although capable of leading normal lives, a split-brain patient will frequently find himself doing unusual things, such as "buttoning his shirt with one hand and unbuttoning it with the other." [17]

Aside from calming seizures in people and totally disabling an unfortunate population of laboratory monkeys, split-brain research has provided dramatic evidence that man possesses at least two, and possibly more, levels of consciousness functioning *simultaneously*. Investigations show that the hemispheres operate semi-independently even when connected. Each "seems to have its own separate and private sensations," notes biologist Roger Sperry, "its own perceptions; its own concepts, and its own impulses to act, with related volitional, cognitive, and learning experiences." [18] Furthermore, each hemisphere tends toward specialization. One side of the brain, usually the left, does better at verbal tasks, while the right hemisphere is more capable at spatial activities such as drawing and designing. Pioneering researcher Joseph Bogen has described the left as "rational, digital, and objective," the right as "emotional, analogical, and subjective." [19]

In our linear, ordered culture, the left hemisphere, or rational brain, clearly dominates our thinking, but its rule is not absolute. At times—during some of our waking hours and perhaps during most of sleep—the right hemisphere, or artistic brain, takes over.

How does this switching process take place? It seems that domination by one hemisphere occurs by default; the adjacent hemisphere simply turns off. Physiologically this means that one side remains in the busy beta state while the other slips down into the rhythmic, less active alpha state. An aesthetic consciousness would indicate that the rational brain is resting in alpha; conversely, a logical consciousness would mean that the artistic brain has shifted to a lower brain-wave frequency.

Further experiments using biofeedback from the rational and artistic brains should yield interesting results, perhaps even destroy some myths. For example, are women really less rational than men? Preliminary evidence already indicates that there is at least more stable hemispheric integration in women than in men. Women are more likely to emerge from left hemisphere surgery with less pronounced speaking difficulties than men. Similarly, "right hemisphere brain surgery seems to impair women's art aptitude less than it does men's." [20]

The big breakthrough, however, will come when man learns through biofeedback training to switch into rational or artistic consciousness at will, simply by turning off the interfering hemisphere of the brain. This is exactly the goal that Bob Ornstein is trying to achieve at the Langley Porter Neuropsychiatric Institute. Is it possible for man to gain voluntary control of his artistic and rational thought processes? "The equipment we need is just at the area of being put together," Dr. Ornstein notes with eager anticipation.[21]

Ornstein believes that learning to generate brain waves

in specific areas of the brain, rather than producing gross alpha or theta, is a quantum jump in brain research. More precise control of brain waves, he predicts, will enable man to master a greater repertoire of moods, emotions and experiences. He further believes that man will discover more than two simultaneous consciousnesses. "I'm sure that's true," he says confidently. "I think we've got many more than two." [22]

Biofeedback Training and Meditation

Psychologists have conducted many bizarre experiments over the last century, but none more unusual than that which took place on April 26, 1956, at the All-India Institute of Mental Health in Bangalore, India. Shri S. R. Khrishna Iyengar, a frail forty-eight-year-old Yogi, walked across the hospital grounds toward the electrophysiology laboratory. He was scantily dressed, wearing only a loincloth and carrying a prayer stick tightly under his arm. Confidently, he stopped above a pit which had just been excavated to rigid specifications by hospital employees. His head was bent, his hands folded in prayer. For years the Yogi had traveled from village to village, demonstrating the powers of meditation to poor farmers and peasants. Eagerly, he now awaited this opportunity to prove scientifically the value of Yoga meditation. After a few seconds he descended into the pit, lit an incense stick and lay down on his back. While the incense slowly burned, laboratory assistants quickly wired the Yogi with electrical instruments so that his vital functions could be monitored. Finally, the pit and its lonely occupant were covered with a wooden plank, which in turn was covered with dirt. Then the experiment began. In order to survive, the Yogi would have to reduce his metabolism enough to sustain himself on minimal air seepage through the dirt.

For over nine hours, a group of psychologists, doctors and medical workers from the World Health Organization waited nervously while monitors attached by wire to the Yogi registered heartbeat, respiration and other life

71

signs. Late in the afternoon, the pit was uncovered. "There on the floor of the pit lay the Yogi exactly in the same position as he had been in the beginning." Immediately, he sat up and got to his feet. Fresh and agile, he accepted garlands of flowers from incredulous villagers who had stood around the pit. The Yogi had actually wanted to stay underground for thirty-six hours, but concerned scientists felt that nine hours was "already much longer than any person could normally live on less than a [cubic meter] of air." [23]

The fantastic powers of meditation are documented beyond a doubt. One swami was measured fibrillating his heart at 300 beats per minute, a maneuver which can be fatal if continued for extended periods. Other recorded feats include slowing the heart rate, raising and lowering temperature in different parts of the body and cleaning the intestinal tract by moving a cloth with the abdominal muscles. Then there is Ramón Torres, a Peruvian meditator who registers high alpha while sticking sharpened bicycle spokes through his face. [24]

Of course, you may not be interested in breathing through dirt for nine hours or jamming the spokes of Johnny's tricycle through your gums. You should, therefore, be aware that meditation has some less spectacular but more practical benefits. Scientists have now established that meditators can reduce mental and physical tension, endure pain more effectively and improve their capacity for empathy with the problems of others. [25] Dr. R. Keith Wallace and Dr. Herbert Benson of Harvard distributed questionnaires to almost two thousand drug users who had practiced transcendental meditation, a Yoga-related technique popularized by the former Beatle guru Maharishi Mahesh Yogi, for at least three months. "It was clear," the scientists reported, "that most were at one point heavily engaged in drug abuse. But practically all of them [95 percent] said that they had given up drugs because they felt that their subjective meditative experience was superior to what they achieved through drugs. And drugs interfered with their ability to meditate." [26] Current clinical researchers are investigating the possible

72

uses of meditation in controlling blood pressure, over-coming addictions (e.g., drug abuse, alcoholism) and reducing anxiety.[27]

These benefits of meditation are certainly important, but as we indicated at the beginning of the chapter the best reason to meditate is to achieve a sense of peace and quiet. The world compartmentalizes us, forcing us to act according to rigid rules and roles. Meditation is a concentrated period of being alone which allows one to resynchronize his internal rhythms. The cortex becomes quiet, allowing the lower areas of the brain to begin percolating in a harmonious rhythm of their own, a rhythm that provides relief from the impingement of the external world for at least a few moments of each day. Meditation literally allows you to get yourself together, to discover an organic center.

Can Biofeedback Training Provide a Shortcut to Satori?

Many Americans might be willing to undertake meditation, particularly if they knew of the benefits involved. Yet many hesitate to do so because they think it would take too long. "Meditation sounds like a good idea," they say, "but doesn't it take years of practice to perfect meditative skill?"

Perhaps not. Meditation, like all other states of consciousness, should have specific physiological correlates in the nervous system. If we find out, for instance, that meditative states are associated with certain brain-wave patterns, we should be able to reproduce these same patterns in a relatively short time through biofeedback training.

Can this really be done? Progress is underway. We already know, for example, that Zen masters produce a predictable series of EEG changes as they move into the meditative state. Two Japanese scientists, Akira Kasamatsu and Tomia Hirai of Tokyo University, were able to classify Zen meditation into four specific stages: the appearance of alpha with eyes open (this is more difficult than with eyes closed), increase in alpha amplitude, decrease of alpha frequency (a further slowing down of the brain waves) and finally the appearance of a "rhythmical theta train." [28] We also know that people who normally produce a well-defined alpha pattern have greater aptitude and enthusiasm for learning certain kinds of meditation.[29]

The viability of biofeedback training as a shortcut to meditation is currently a hotly debated issue among researchers. Purists assert that no amount of brain-wave training can duplicate the meditative state. They argue that Zen, the various paths of Yoga, transcendental meditation and so on are states of being that go beyond mere physical description. On the other hand, many people already have claimed to have achieved meditativelike states through biofeedback training, particularly alpha training. Dr. Johann Stoyva, an outstanding researcher in the voluntary control of the nervous system, has suggested that information feedback techniques might be able to teach the "blank mind" state, typical of Zen and Yoga, within "months or even weeks." [30] (Stoyva's own experience of alpha was "like a flowing grey-black film with a luminous quality.") Kamiya, too, believes that biofeedback will be useful for producing meditative states, although he does not think the answer lies strictly in the alpha frequencies: "I do think . . . it will be possible to find the unique neurophysiological signature of meditation by checking out other channels—besides alpha. Once we have the complete physiological pattern that characterizes meditation, there's no reason why we can't train people, with feedback, to mimic it in a relatively short space of time." [31]

The debate between those who see biofeedback training as a shortcut to meditation and those who do not is likely to be resolved in the near future. As the technology for measuring and training brain waves becomes more sophisticated, unpracticed meditators will have the opportunity to duplicate the physiological states of Zen and Yoga. Of course, even if scientists are able to produce the meditative consciousness together with its concomitant psychological benefits, the purist could still argue that biofeedback training is not the "real thing." Then again, if voluntary control of brain waves can simulate all the advantages of meditation, who cares?

Using Biofeedback Training to Achieve a Calm State of Consciousness

I felt I was floating above the chair.

It sounds funny, but . . . well, okay . . . it seems like there was some kind of force on the inside, flowing through my forehead . . . not a hard pressure but you can feel it, like when you move your hand through flowing water.

I'm not even sitting here. I feel like I'm just detached in some way . . . you know, if I create some sort of image I feel as though I'm just there.

These are some of the actual statements made by subjects describing their experiences in a unique experimental program which began at the Menninger Foundation in the late 1960s.[32] The purpose of the program, according to researchers Elmer and Alyce Green, was to compare biofeedback training and autogenic training.

Autogenic what? Unfortunately, because of academic psychology's traditional bias against concepts such as "voluntary" and "willpower," most Americans have never heard of this valuable therapeutic technique. Autogenic training is essentially a self-induced method for achieving inner calm and relaxation which incorporates aspects of hypnosis, psychoanalysis and Yoga. Developed for clinical use in 1910 by a German named Johannes Schultz, it is widely practiced throughout Europe.

In one experiment, which the Greens appropriately dubbed the "triple training program," the goal was to enhance the calming effectiveness of autogenic training with biofeedback training. Volunteers were taught to perform three tasks simultaneously: reduce muscle tension in the right forearm, increase the percentage of low-voltage alpha rhythms in the brain, and increase temperature in the right hand (an indicator of body relaxation).

75

At first inspection the project would seem to be a clinical nightmare, but in practice the procedure was both elegant and rewarding. The experimental situation itself looked like the familiar TV simulation of an astronaut in a space capsule. The typical volunteer, wearing a special research jacket equipped to gauge respiration, sat comfortably in front of a screen panel. Electrodes attached to the back of his head, two fingers of his hand and his right forearm transmitted brain- and body-wave feedback to three bars of light on the screen, each bar growing taller or shorter in accordance with specific body changes. For example, as the subject reduced muscle tension, the first bar would rise. The second and third bars rose and fell with respective increases and decreases in hand temperature and percentage of alpha rhythm.

The training consisted of subjects repeating autogenic phrases while, at the same time, receiving biofeedback from the three bars. Typical autogenic phrases were: "I feel quite quiet," "My hands are heavy and warm" and "My ankles, my knees, my hips feel heavy and relaxed." The subject's only goal was to keep the bars as high as possible through "mental powers."

The results were striking. After testing sixty subjects during a period of over three years, Dr. Green concluded that triple training achieves a relaxed consciousness far sooner than autogenic phrases by themselves. In one study of muscle tension alone, "seven of 21 subjects achieved either complete or near complete relaxation of striate forearm muscles in less than 20 minutes of a single lesson." [33] To become proficient at relaxation through autogenic training by itself normally takes four to six months.

The implications for daily living are obvious. After a few training sessions on the machines, chronically tense people could learn to alter their states of consciousness—and learn to relax themselves quickly and at will. Psychiatrists would also find this technique useful for therapy. Patients who could learn to induce a calm state of consciousness would be able to cope with severe anxiety-producing problems more effectively.

In the beginning, most brain-wave research centered on alpha. There were obvious reasons for this. Alpha was (and is) a relatively easy frequency to monitor and control. In addition, the relationship between alpha and experiences of pleasure was an incentive to imaginative scientists. Recent evidence, however, indicates that theta is a far more intriguing brain-wave state. Zen masters, for example, go *past* alpha into theta during deep meditation. Theta is also the state people reach just before sleep, a time when writers and artists usually get some of their best ideas. (Many creative people habitually keep a pencil and paper near their bedside in order to record the insights that occur during pre-sleep theta states.)

Dr. Green first became aware of "theta power" when, after the triple training program, several subjects who had produced low alpha and theta waves reported states of semiconscious reverie punctuated by hypnagogic and dreamlike images.[34] Subsequent experiments have convinced Dr. Green that people who travel into theta are able to experience psychic events that are normally buried in the unconscious mind. He believes that theta training may be very useful for analytic therapy. "Psychiatrists will be able to develop in many patients a deep reverie in a short period of time through the use of feedback techniques for deep relaxation [and] with selected cases . . . normally unconscious material of analytic value should be recoverable."[35] He also believes that theta training will enable man to explore higher levels of awareness, including creative consciousness, without the use of drugs.

What is theta training like? One woman who participated in a three-month program at Dr. Green's lab describes it this way: Each day she would go into a darkened, empty room, casually place electrodes on her head and then plug the wires into an EEG feedback circuit. Guided only by the cool, dim lights from the surrounding machines, she made her way to a nearby couch. "I would attempt to get into a quiet, relaxed, blank mind state, and yet not fall asleep," she recalls with a smile. "And that's not easy to do!"[36] She would proceed to stretch out, calm herself and wait for a tone to indicate that she was in theta. Her task was to become aware of her experience while the tone was sounding. "Fifty percent of the time . . . I couldn't pull into consciousness what was just under the level of consciousness, and I probably would have gone to sleep if it weren't for the tone." Other times, however, she experienced sensations and images "so very personal that I wouldn't want to discuss it except with a psychiatrist or guru." The material she would discuss went something like this:

"I saw fantastic images. . . . There was an American Indian with a feather headdress on a long refectory table. . . . There were some images that were reminiscent of my own nighttime dreams. . . . There was on a couple of occasions an eye, just a very large eye . . . pulsating with light and life. . . . There were also differences in body sensations . . . feeling my body rising as if a tingling, buzzing from inside the core of my limbs was making [me] rise from the table. . . ."

Somewhere between waking consciousness and sleep lies a wealth of unexplored images and experiences. The limits and value of this theta state are only beginning to be explored.

Hypnotic Breakthrough

Dr. R. August wanted to avoid fetal narcosis.[37]

His patient, a twenty-one-year-old girl from Muskegon, Michigan, named R. C., was in premature labor after six months' gestation, and anesthetics might induce a trauma in the fetus. The solution: using hypnosis during delivery to divert the girl into reliving a pleasant past experience and thereby avoid pain. Dr. August knew that R. C. enjoyed traveling by car and that she had made trips south to Chicago and north to the Straits of Mackinac. "As she expressed no preference for either," he wrote later, "I chose the journey south because of my greater familiarity

78

with this route. I aided her in hallucinating her husband driving the car and herself sitting in the front seat watching the trees, telephone poles, houses and the cars passing in another lane."

As the delivery progressed, the doctor decided to see how the journey was going and was surprised to discover that she had not even reached Grand Haven, just a few miles south of Muskegon. "Where are you now?" he asked. "Ludington," she replied, referring to a northern town. "Subconsciously she wanted to drive north but had been considerate enough to permit me the choice," explains Dr. August. "I turned around to accompany her. We enjoyed the journey together but reached the strait bridge before the delivery. So I asked her husband to drive very slowly as we wished to notice the sky, the water, the boats, and other cars."

The baby finally came. It was two pounds seven and a half ounces, but healthy. "[R. C.] later told me that the bridge crossing was the slowest trip she had ever taken," the doctor notes.

Except for a few unexpected turns, hypnosis is a safe and reliable process when conducted by a competent physician. Although scientists will debate over the exact physiological nature of hypnotic consciousness,[38] there is no longer any doubt about the effectiveness of the technique. Dr. August uses hypnoanesthesia in over 80 percent of his obstetric practice.[39] Experiments too numerous to elaborate have demonstrated the power of hypnosis to induce or ameliorate myriad medical disorders [40] including pain, warts, allergic skin reactions, uncontrollable urination and cold stress. Equally as important, the hypnotic state is characterized by a level of tranquillity and inner peace that is far above normal waking consciousness.[41]

Until recently, hypnosis did not receive much attention from the medical and psychological professions. Despite its obvious clinical advantages, hypnosis had its limitations. Freud, for example, gave it up when he discovered that not everyone could be hypnotized. Other scientists have ignored hypnosis for the same reason. Recently Dr. Joe Hart lamented, "We have reliable measures of hypnotic susceptibility but . . . no ways to change a person's responsiveness markedly to hypnotic instructions." [42]

The breakthrough for Dr. Hart and his colleagues came through two experimental studies. In the first one [43] over fifty subjects were examined with EEG machines to determine if hypnotic susceptibility was related to specific brain-wave states. The mathematics was complex, but the results were rewarding. There *was* a relationship. "The most productive EEG frequencies were from the slow frequency range," wrote Hart in a learned scientific journal. In other words, the key was alpha.

The next question was obvious: Could hypnotic susceptibility be improved through alpha training? The answer came in a second experiment,[44] this one involving thirty volunteers recruited through newspaper ads. Again the math was complex, the results rewarding. And again, the results were reported in distinctly unspectacular academic tones: "hypnotic susceptibility and operant alpha rhythm are positively related both before and after subjects have received alpha training." [45]

The ultimate impact of these experiments could be monumental, though it is too soon to make specific predictions. Certainly, more people will have access to the medical and emotional benefits of hypnosis. Also, a combination of alpha-hypnosis training may enable each of us to develop our general capacity for brain-wave control and what Hart calls "mental fluency."

A Little Red Schoolhouse for Consciousness

2 P.M., Wednesday. It is a typical day at school. Jim is in the upstairs bedroom making final adjustments on the EEG machine. Carefully he wipes the silver disk electrode with alcohol, then covers it with a whitish paste. Skillfully parting his sun-bleached hair, he places the electrode directly against the scalp at the back of his head. Once more he checks the controls; then a flick of the switch and the room is flooded with harsh static. Jim reaches for the correct knob and turns it slowly until the screech dissolves into a soft rhythm of purrlike clicks. Breathing

deeply, he falls gracefully into an oversized leather chair. Simultaneously, the gentle clicking sound perks into a muffled roar. Jim smiles. After six sessions, he knows he has mastered the alpha rhythm.

From Jim's bedroom window you can see the large front lawn. A frisky Labrador retriever is staring impatiently at his mistress, who lies motionless on the soft grass. Looking at the small box in her hand and the headphones over her ears, you might think she was listening to the quiet strains of a Mendelssohn concerto on FM radio. You might, were it not for the band of electrodes pasted to her forehead. Music may soothe the jangled nerve, but she has found EMG feedback more relaxing. Today she is learning to gain control over the frontalis muscle just above her eyes. With a little more practice, she may never worry about tension headaches again.

Directly on the other side of the school building is another lawn, smaller but better shaded, where Rob is conducting an informal class discussion of biofeedback training experiences. His students include a college teacher, a rancher, a clergyman, a law student, a writer and an itinerant motorcyclist. The reports of training experiences are as varied as the people themselves. "I felt turned on physically in the lower part of my body on my first try with alpha," volunteers one student. "For me it was a funny floating sensation," says another.

Rob smiles. His years as a biofeedback scientist have taught him that the range of human experience is unlimited. Despite his prestigious Ph.D. and other impressive credentials, Rob knows that his knowledge of human consciousness is still at the novice level. Carefully he listens to each student, trying as hard to learn as he does to guide.

Mike would also like to learn about his fellow students' alpha experiences, but he is into another trip right now. Reluctantly, he closes the large center window, shutting out the noise of Rob's class from the spacious downstairs living room of the school. Quietly, he returns to a group of friends sitting on the floor in a circle around a tall slow-burning candle.

Then the meditation begins. One minute staring at the

candle, the next minute trying to visualize the flame with eyes closed. Again, one minute staring at the candle, another with eyes closed. The process is repeated a third and final time; then the group relaxes and begins exchanging individual accounts of the experience. A few people want to continue the exercise, but Mike knows better. This form of meditation is best performed with moderation. Better to leave it for now and get into some simple yoga positions.

These scenes are not speculations on what a future biofeedback training school might be like; they are actual events that took place at Stanford University during the summer of 1970. One of the authors, along with over thirty other "students," took part in the Prana/Esalen Institute-sponsored summer school on the Psychology of Human Consciousness. The program, which lasted over two months, attempted to integrate courses in Yoga, Eastern religions, body therapies, Gestalt psychology, history of consciousness and, of course, biofeedback training.

The focus of the program was not philosophy but individual development. Each person was encouraged to explore new realms of consciousness and to apply his discoveries in practical ways that would alter his life. In many cases, the results were dramatic. One bright young man decided to leave his life as a recluse and start a school dedicated to furthering human awareness. Another gave up a lucrative academic career in order to explore biofeedback. The motorcyclist decided to retire to a four-wheeled van.

To what extent was the summer school really responsible for these changes? Will consciousness-expanding programs such as this one ultimately become as much a part of standard education as courses in reading and writing? At this point these questions are unanswerable. But one thing is clear: Consciousness-control programs using biofeedback training and related techniques are already with us and growing. Their ultimate success is limited only by the level of man's technological skill and the boundaries of his imagination.

What to Do till
the Revolution Comes

Above all, become a connoisseur of your own nervous system.

Let's presume that having read this far in the book you're excited enough to want a taste of biofeedback training yourself. Where do you turn? "In five years," predicts Dr. Barbara Brown, "there will be biofeedback centers all over the country, in which people can learn all manner of mind and body functions." [1] But that's still a few years away. Where do you go until the revolution comes?

There are several alternatives available. Some of these can be beneficial if properly exercised; others, quite frankly, can do little more than waste your time and money. It will be our task to discuss and evaluate these various alternatives in this chapter.

Purchasing a Biofeedback Machine

An excellent way to get biofeedback training would be to go out, buy an EMG or eight-channel EEG machine and hire a medical team to give you proper supervision. The only problem is the cost: a laboratory EEG machine runs $10,000, not including the medical team.

An alternative strategy is to go out, purchase yourself a portable biofeedback machine and supervise your own training—at an approximate savings of $9,800. With that kind of bargain, who could afford not to have a go at biofeedback training? Hardly anybody, it would seem, judging from the profusion of new companies who are flooding the market with low-priced machines. These little metal boxes blip, hum and bleep in all imaginable ways. One even announces alpha with a birdlike warble. Some are equipped with gold-plated electrodes; others come with lanolin-enriched electrode cream, flashing lights and grain leather cases. The only question is: Do they work?

That is the $200 question. The answer is a *qualified* yes. It depends on several factors, including the quality of the machine and how it is used.

There are three types of biofeedback machines currently available on the market:[2] the EEG machine for controlling brain waves, the EMG machine for controlling muscle tension, and the temperature-feedback machine for controlling blood flow in different parts of the body. Let us consider each of these machines individually.

EEG Machines

The EEG machine, as we have indicated throughout the book, is used for monitoring and controlling brain waves. Most commercial models resemble portable tape recorders, come with three electrodes which are easily attached to three points of the head, and are designed primarily for alpha brain-wave training. A few machines record both alpha and theta and are more expensive. Prices vary from $100 to $700, with a median cost of about $200. Although cost is usually an indicator of quality, it is not an absolute guarantee that the machine will work effectively—that it will accurately sense your alpha waves and, in turn, give you correct feedback.

The EEG machine is an extremely delicate and complex instrument; the number of its components are equivalent to the parts in a score of radios. Elmer Green has tested a wide variety of commercial EEG machines and finds that "even the most expensive don't always work right." [3]

But even a perfectly calibrated portable EEG machine is not enough for successful monitoring and control of your alpha rhythms. *You must have proper instruction.* As Dr. Solomon Steiner of the City University of New York points out, a wide variety of facial expressions, or "artifacts," as he calls them, can pass for increased alpha. They include eye blinks, twitches, tensing of the muscles in the forehead, raising your eyeballs, frowning and even gritting your teeth. Someone who thinks he is producing alpha might be doing nothing more than reinforcing jittery eyes. [4]

For a person wishing to train his brain waves immediately, we suggest the following procedure. First, call a local university to see if any professors are conducting alpha, theta or delta biofeedback research. If so, volunteer to become a subject. You may be able to get on the best machines under expert supervision for nothing. The experimenter might even pay you for your time.

If, however, university training is not available, it will be necessary to purchase a machine of your own. In this event, we recommend that you buy one of the higher-quality inexpensive machines. As we have indicated, they are not guaranteed to be perfect, but on the whole they *will* give you better performance. The following companies produce what we consider to be, as of this writing, acceptable lower-cost feedback machines.

Cambridge Cyborgs Corporation
4 Brattle Street
Cambridge, Massachusetts 02138

New York Cyborgs [5]
26 West 9th Street, Suite 9C
New York, N.Y. 10011

Scott Behavioral Electronics, Inc.[6]
Box 3306
Lawrence, Kansas 66044

Toomim Bio-Feedback Laboratories, Inc.
10480 Santa Monica Blvd., Suite 1
Los Angeles, California 90025

Toomim Bio-Feedback Laboratories, Inc.
41 West 71st Street, Suite 1C
New York, N.Y. 10023

Once you have selected your machine (each company will send you cost and performance specifications upon request), be sure to get proper guidance in how to use it. To guarantee this, make your purchase contingent upon the receipt of adequate personalized instruction. (Some prospective customers prefer to take these training sessions *before* they purchase the machine, so they can see if they want to buy it. This alternative is becoming more feasible as additional companies rent their machines for an hourly fee—usually around $10.) If you live too far from the area where the company is based or has an office, ask the firm to refer you to a competent EEG instructor near your home. Because brain-wave feedback can be confused with facial artifacts, *we strongly recommend that you do not use a machine without professional guidance*.

Instruction by a company is conducted in small groups or in individual sessions. There is a trade-off here: group sessions usually cost less, but you learn more quickly with individual attention. How long should training take? There is no set answer. Individuals vary in the time it takes them to get the hang of their machine and the alpha (or theta) experience. In scientific laboratories it usually takes subjects anywhere between four and ten hours to learn alpha brain-wave control. (Some subjects, however, don't seem to produce any alpha at all.) We also think it advisable for individuals, once they have mastered their machines, to return for periodic training checkups to make sure they haven't developed any bad habits in the interim.

If at all possible, we also recommend that you check your machine's accuracy on a signal generator or oscilloscope built for that purpose. Usually this equipment will be available at a university or medical facility. If this seems like a superfluous effort, it isn't. Today the consumer interested in biofeedback training is in the same position as those people who wanted to buy the first automobiles. The machinery was expensive, unfamiliar and not always reliable. Like driving an automobile, biofeedback training will require a few years to become a simple and common part of American life. When it does, Dr. Steiner envisions biofeedback centers set up along the lines of auto diagnostic centers where people could come in with their inexpensive machines and check them out against the readings of $10,000 models. At the same time the owner "could practice with the good equipment to learn how to properly use his inexpensive machine." [7]

Finally, don't anticipate that brain-wave biofeedback training will necessarily be ecstasy. As we explained in Chapter 3, the experience of alpha and theta is not the same in every person. You will learn more about yourself and gain control of your brain waves more quickly if you come to EEG training without preconceptions of the experience.

Before going further we should add some words of caution. Many biofeedback researchers are divided in their opinions on the benefits of brain-wave biofeedback training. Dr. Eleanor Criswell at Sonoma State College takes the very liberal position that using biofeedback machines is good for you, even if the machinery is faulty and doesn't give you control over your brain waves. "It doesn't really matter in the sense that if [people are] seated and quiet, if they are listening to the tone and auditory feedback . . . they probably can put themselves into a meditative state using the device." [8] Other investigators laud brain-wave training, pointing out that alpha is associated with states of pleasant relaxation while theta seems positively associated with creativity. Yet a few experimental findings give us reason to be cautious. For a small minority of the population, alpha is experienced as an anxious state. The

87

psychiatric reasons for this are not clear. Some scientists believe that alpha is always pleasurable, but that a few puritanical people are conditioned to experience anxiety at the prospect of pleasure. Others simply say that the alpha experience varies from person to person. Still others believe that some alpha configurations are pleasurable, others are not, and that the present machinery is not sophisticated enough to record the differences.

Then there is the possibility that alpha might be correlated with underachievement.[9] Using an advanced computer technique, one scientist discovered that students who do poorly at school also evidenced abnormally high alpha-wave activity. Does this mean that biofeedback training with alpha will make a person less motivated? We do not know, but the possibility must certainly be investigated.

Theta, too, is occasionally accompanied by unpleasant feelings. Biofeedback engineer Herschel Toomin speaks of an experiment where twenty-six people were trained to go into theta. Although half the subjects reported experiences that could be called creative, a few actually experienced frustration, tenseness and anxiety.[10] Swami Rama also describes theta as a disturbing state where all "obligations and demands" come into consciousness. Why should theta be experienced as both creative and unpleasant in some people? Does creativity require anxiety? Again, no one really knows.

Where do these different opinions leave us? Our position is that many people will find brain-wave training beneficial, but that the present advantages of alpha and theta training are probably exaggerated by those people who sell low-priced biofeedback machines. This is particularly true when you consider the state of the art; none of the consumer-oriented machines, for example, give you the ability to control alpha or theta in specific parts of the brain, an ability which promises to make brain-wave control infinitely more valuable. On the other hand, we do not consider brain-wave training dangerous provided you follow the guidelines provided in this section and, of course, terminate training if you experience any psychological or physical discomfort. (At present there is no medical evidence to suggest that brain-wave training is harmful in any way.) The worst that could happen is that, in the process of listening to your brain, you become relaxed and quiet. When you think about it, that isn't so bad.

Before leaving the topic of brain-wave machines, a final observation is in order. Recently, we have come across an advertisement for an alpha-type machine which, instead of simply monitoring a person's own brain-wave activity, *actually delivers a small electrical pulse to the head. Theoretically,* this encourages the brain to generate its own alpha waves. We emphasize the word "theoretically." Scientifically, no one knows if this machine has any value at all. Until more research becomes available we recommend that this machine not be used. We might also mention that such an instrument flies in the face of biofeedback philosophy, which emphasizes the person's ability to control his inner states by his own mental power, not some machine's electric power.[11]

EMG Machines

Although most of the popular literature on biofeedback centers on brain-wave training, body-muscle training with EMG feedback machines appears to hold more promise medically and even for controlling mental states. Originally, researchers saw alpha training as a major weapon to combat depression, anxiety and other mental disorders. Recent evidence, however, indicates that alpha training has some shortcomings. For one, alpha is not always a sign of reduced anxiety. Some people can be perfectly relaxed yet show no alpha readings. Budzynski and Stoyva believe that EMG biofeedback, which goes directly to the muscle core of anxiety, is a more promising area of clinical investigation.[12]

What about the reader who wants to purchase an EMG machine? It won't be as easy as purchasing a brain-wave instrument. First of all, not as many companies manufacture EMG machines—although this condition is changing. Secondly, the good EMG machines which are produced are often more expensive than EEG instruments; the one we recommend, for example, costs $450. Finally, and most important, most of the legitimate high-quality companies will not (or do not want to) sell their instruments directly to the public. They would rather sell to doctors, psychiatrists, clinicians or other responsible professionals and let the public use the machines under their trained supervision. There is good reason for this: improper use of EMG equipment can be dangerous either because the buyer uses the machine in an incorrect way (for example, intending to monitor and regulate muscle tension in both arms a person might erroneously monitor and alter his heartbeat) or for an incorrect reason (treating his bodily symptoms without medical supervision).

Let us elaborate on this problem of self-diagnosis a bit further. Biofeedback training is a marvelous tool for controlling bodily malfunctions *as long as you know what these malfunctions really are*. For example, a pain in the cranium is not always a muscle-tension headache; it can also be a brain tumor. Trying to treat a tumor with an EMG feedback machine would not only cost you money and time; it could also cost you your life. The point is this. Playing with brain-wave biofeedback machines to achieve altered states of consciousness is fine. However, before attempting to cure symptoms with biofeedback training— be they headaches, cold feet or anxiety—be sure you know what the symptoms actually signify. *Consult with a doctor first!* By all means, become your own prescription for good health, but only when the illness is correctly diagnosed.

If you wish to experience EMG biofeedback training, we suggest you call your doctor and see if he or anyone he knows has such an instrument. You might also try your local university (as we explained with brain-wave biofeedback). If you can't locate an EMG machine in your area and you would still like to buy and try one, we recommend you order the machine through your physician and let him train you on it. The machine we recommend is the portable EMG feedback system, Model PE-2 ($450), manufactured by:

Bio-Feedback Systems, Inc.
P.O. Box 1827
Boulder, Colorado 80302

This model is one of the machines used by Budzynski and Stoyva in their muscle-relaxation work (tension headaches, anxiety reduction, etc.). The machine provides feedback that is both auditory (tone through a headset) and visual (through a meter) and comes with a three-electrode band that can be conveniently worn on the forehead.

Temperature Feedback Machines

What we have said concerning EMG machines also applies to temperature feedback instruments. "Temperature trainers," which have already shown promise in the treatment of migraine headaches, Raynaud's disease and other medical disorders, should only be used under a doctor's or scientist's supervision. The machine we recommend is the T1-A model ($160) produced by:

Scott Behavioral Electronics, Inc.
Box 3306
Lawrence, Kansas 66044

This instrument is manufactured under license from the Menninger Foundation and has been used by Dr. Green and Dr. Sargent in their research work. It consists of two temperature-sensitive electrodes (one worn on the middle finger of the hand, the other on the forehead) and a meter which registers the temperature difference between the subject's head and hand.

A Call for Regulation

As biofeedback training becomes a big business, reputable companies and scientists alike are coming to realize the pressing need for immediate impartial evaluation and control of machines on the market. Regulation is particularly crucial, especially as more "medical" biofeedback machines (like blood-pressure and EKG units) are produced and allowed unrestricted use. We are coming to understand that man is an exploring creature: give him a chance to use a new machine and he will do so—sometimes, through ignorance, unwisely. We are also coming to appreciate that man has awesome power to regulate his nervous system—and to abuse it. Some of Neal Miller's rats have died while slowing their heartbeats to low levels.[13] Of course, rats are not humans, and these particular animals were also paralyzed with curare. Yet we know that man can alter his heartbeat significantly. Might a person who has an undiagnosed case of heart disease damage himself through self-imposed biofeedback training? We don't know—and until we do we must proceed carefully.

One of the most encouraging experiences the authors had while writing this book was talking with John Picchiottino, president of Bio-Feedback Systems, and Kenneth Scott, president of Scott Behavioral Electronics. Both these men are in business and must be concerned with such things as profit and loss, yet both told us they didn't want to open their market to the public at this time. Mr. Scott put it this way: "These instruments in the hands of an amateur could be just as dangerous as drugs and should be regulated." [14] We agree. If a businessman is willing to give up business to ensure the safety of the buying public, then we, as consumers, and the government, as our watchdog, should be willing to follow his suggestion.

Instant Nerve-ana

Two kinds of organizations are capitalizing on biofeedback research. Those that we mentioned at the beginning of the chapter actually manufacture, sell and instruct people in the use of biofeedback instruments. Those in the second group simply use terminology from biofeedback research as an advertising gimmick to sell their own brands of "instant nerve-ana," with little or no use of actual biofeedback equipment. Without some careful inquiries, it is almost impossible to tell which organizations actually use biofeedback equipment and which don't. This is because the "instant nerve-ana" groups word their advertisements to imply as close a connection with biofeedback as legally allowable. Of course, once you actually begin participation in a "biofeedback seminar" it will become readily apparent whether your chosen organization does or doesn't use biofeedback equipment. The problem is that most organizations, whether or not they use instruments, require all or part of the course fee in advance. Thus one must know *before* he registers for a course whether or not it will include the use of biofeedback equipment. Here is a suggested checklist procedure that will help you make this determination.

1. If you see an advertisement for brain-wave or mind control, write the company a letter requesting the following information (or call, ask the questions below and demand a written reply):

Dear Sir:

I understand that you are offering a course in mind control. In order to decide whether I should enroll in your course I will need answers, in writing, to the following questions: (a) If I become a participant in your program will I have a chance to use (not just see demonstrated) biofeedback equipment? If the answer to this inquiry is "yes," (b) What kind of instruments (EEG, EMG, etc.) will I get a chance to use and (c) about how long will I actually be using this equipment? (d) (optional question) Could you please list the manufacturer of the biofeedback equipment I will be using? [15] Could you also be so kind as to send along any brochures describing your program?

Thank you.

2. If you get no answer to your letter or call, or a reply indicating that machines are *not* utilized, we recommend

that you do not enroll in the program (assuming, of course, that you want biofeedback training).

3. If you get an answer to your letter indicating that machines are used, that they are the kinds you are interested in (if you want brain-wave training, don't sign up for a program advertising EMG biofeedback), and that you will get a chance to use them for a reasonable amount of time, you might want to consider enrolling in the program.

4. Call your local Better Business Bureau to see if any complaints have been filed against the organization.

5. If the company has enclosed any brochures describing their program, look for the names and qualifications of those associated with the organization. Do they have a board of advisers? If so, are they Hollywood movie stars or are they scientists? What about the actual teachers? Do they have any professional degrees (Ph.D., M.D.)? An organization doesn't have to have a professional staff to be competent, but it helps!

6. If, after all these steps, you want to enroll in the program, call the organization and ask if a free (or low-cost) preview or demonstration class will be held prior to registration. If so, attend and see what you think.

Although these six steps might sound cut and dried and easy to follow, beware! Even if you find out that a given organization doesn't use biofeedback equipment (or uses it too sparingly to be worthwhile), it might still be difficult to resist the temptation to join. This is because many of the "instant nerve-ana" companies make their ads and demonstrations so appealing it is difficult to say "no." There is nothing illegal in this approach; they have a product to sell and it is reasonable to assume they will package it as attractively as possible. It is up to the consumer to be a discriminating judge of the advertising. Again, this will not be easy. How many of us, for example, wouldn't pay to learn a mind-control method for better memory, better health, better sleep, better attitude, better learning ability, better self-image, better time management, better intuition, more success, more happiness, more self-confidence, more creativity, more energy and vitality, more productivity, more capacity and more friends. Impossible? Silva Mind Control doesn't think so. In a recent ad they claimed that people who learned their mind-control method could accomplish such things. Who would believe such claims, you ask? In the same ad, Silva Mind Control claimed to have over 10,000 graduates who have completed their course.

Sometimes an organization's demonstrations can be even more spectacular than its ads. One of the authors had the opportunity to attend a series of lectures billed as a preview of "Mind Dynamics." The first one was held before three thousand people who crowded into the convention hall of the Jack Tarr Hotel to hear men like Bob Cummings and U. S. Anderson (no kidding—he even wore a red, white and blue tie) sing the praises of Mind Dynamics. They were followed to the podium by Alexander Everett, president of Mind Dynamics. Making vague references to biofeedback training and other advances in brain-wave research, he explained how the average man uses only a small fraction of his brain and how, by learning to tap the "hidden potential" of the mind, each member of the audience could achieve, fame, fortune and every other conceivable goal. The key to unlocking these powers, he revealed, is Mind Dynamics training. "But no more for now," he stated. People wishing further information were instructed to attend one of several upcoming "in-depth" seminars.

A week later a second meeting was held. Approximately half of those attending were graduates of the Mind Dynamics training program; the other half were new recruits. The meeting began with four graduates proudly introducing their guests. A woman named M was congratulated for bringing a man she met on a streetcar while coming to the meeting. Then came the testimonials.

One by one, the grads stood up and told how Mind Dynamics had changed their lives. "I've learned to understand myself more," claimed one young girl, "to relate better to others, to be more positive." A man stated confidently, "Mind Dynamics for me is life. Each day, through

Mind Dynamics, I find one more way to be more alive."
Another young girl insisted joyfully, "Besides finding that
you can have anything you want, that you are the reason
for everything, you also find that you can't be sad or
depressed anymore." A few more such assertions, and
the guests were eager to share in the marvels of Mind
Dynamics.

The hostess, Ms. C, made the final sales pitch. After
detailing how Mind Dynamics had helped her find a meaningful life, she described how Alexander Everett had spent
years synthesizing the most important material from different religions and philosophies into one system: Mind
Dynamics. There are now fourteen instructors in the San
Francisco Bay Area alone, she stated proudly, over a
hundred throughout the United States and the Common-
wealth countries. Her description of the actual techniques
used in Mind Dynamics was vague: it is a thirty-six-hour
program which teaches students methods for "releasing
psychological tension," for "triggering alpha" to guide
their lives, "for tapping unconscious levels beyond time
and space." The price, however, is not vague: $200 for
two weekend training sessions, plus follow-up courses in
the evening. Tuition also includes the option to repeat the
course as many times as needed. (This price is exceed-
ingly high for any organization offering weekend libera-
tion. The prestigious Esalen Institute, for example, rarely
charges over $100 for a comparable time period, although
to be fair, Esalen does not allow free reruns.)

Many journalists and scientists have dismissed Mind
Dynamics and its sister organizations (e.g., Silva Mind
Control) as charlatans. Dr. Kamiya has complained about
the way his name is used to promote these companies, and
Dr. Elmer Green has gone as far as to debate representatives of one "mind-control" organization on television.
Given the hoopla approach of these groups, blanket criticism is understandable. And yet to dismiss them out of
hand, we feel, would be a mistake. Many leaders of these
companies are sincere, dedicated people. Everett, for example, says he does not really enjoy the kinds of flam-
boyant promotional meetings—"circus shows," in his

words—that we have just described but considers them necessary to attract attention.[16] He is also aware that people may gravitate to Mind Dynamics for the wrong reasons. Consequently, he makes more than token efforts to weed them out. We can even sympathize with his views of conventional self-help methods, psychiatry in particular, as being too long and too costly. And then there is anecdotal evidence that for some individuals Mind Dynamics does seem to provide some benefits.

Yet the important question to ask is not "Do the techniques of these 'mind-control' organizations really work?" but rather "How beneficial are they in comparison with other methods of psychotherapy and self-control?" In this context, Dr. Green provides the most intelligent and damning criticism. Most of these companies, he points out, use nothing more than variations on hypnosis. "What bothers me about the whole thing," Green states, "is that I am much more in favor of voluntary control than . . . hypnotic control. [Students in these organizations] go through a four-day program of intense hypnotic education in order to do the things they demonstrate." [17]

For the most part, we agree with Green. As we indicated in an earlier chapter, hypnosis has great medical value when used by a competent physician, but to allow your motives and attitudes to be controlled by an external authority is a questionable prescription for self-advancement. To make matters worse, the precise methods used by instant salvation groups vary constantly and cannot be thoroughly evaluated.

It is not our intention to brand all "mind-control" groups that do not use biofeedback devices as charlatans. On the contrary, the field of voluntary control of behavior is sufficiently new to welcome creative experiments from all interested parties. But we do feel that some kind of screening, perhaps from the FDA or the FTC, is necessary to ensure both the public welfare and the reputation of biofeedback research. Those companiees that honestly represent their services—and have something to offer—would have nothing to fear from such an action.

A natural complement to biofeedback training is body charting, a simple and valuable method for self-control. But to understand how this works requires some further insights into the nature of how humans function.

Most people think of the body as a soft machine, a kind of "gushy" Chevrolet. You take in food-fuel, wash it down the mouth-gas tank into the stomach-engine, where it combusts, giving off nutrition-energy. Weight is considered to be an uncomfortable bulge in the super-structure; indigestion is simply the metal corroding. The body is acknowledged to grow and shrivel with age, but this is a long-term process that does not alter our view of ourselves as hulks.

In fact, the body is quite different from this stereotype. The human organism is a complex of interrelated, clock-like rhythms. These patterns vary so dramatically that a given body at twelve noon may be in quite a different physical state just a few hours later. "Most people don't realize how much they change every 24 hours," states science writer Gay Luce. "They may notice that they get particularly tired at 2 A.M., or chilly in the late evening. [However], they remain largely unaware of changing immunity to infection or stress (which drops at night) or the fact that blood pressure, mood, pulse, respiration, blood-sugar levels . . . and our ability to handle drugs all rise and fall in a circadian rhythm."[19]

The obvious rhythms are those that parallel the twenty-four-hour cycle. For most of us, changes in temperature, hormone levels and alertness occur at predictable times during the day. However, there are biological rhythms of shorter and longer duration. Breathing through the nose, for example, is an involuntary shifting back and forth between left and right nostrils at three-hour periods. The menstrual cycle, on the other hand, is usually a twenty-eight- to thirty-five-day "period." (Men have less known monthly cycles, usually slight variations in body weight.) Mood changes, when averaged out, will show cycles as long as three to six months. Even childbirth follows a pattern, as most babies are born between midnight and 6 A.M.

Body charting is a method for keeping track of your hourly, daily, monthly and even yearly body rhythms. If, for example, you note at what periods of the day you feel energetic and at what times you feel drowsy, a pattern will eventually emerge, indicating your daily alertness cycle. Effective body charting should embrace a period of three to six months, but pronounced cycles will often become obvious within a few days.

The purpose of charting your body cycles is simply this: by learning what your rhythms are, you can plan your activities to best advantage. Some women have already done this by using the rhythm method of birth control, whereby pregnancy is avoided or initiated by a daily charting of body temperature fluctuation in order to pinpoint ovulation (this method is not foolproof, however). If you discovered that your peak alertness period is in the afternoon or evening, you could arrange to do work that demands concentration in the later part of the day. If, on the other hand, you turn out to be an early riser, you could arrange to begin working on important tasks as soon as you get out of bed. By charting your daily and monthly mood rhythms, you could prepare in advance for low points on the cycle. This would be particularly helpful for businessmen. Important conferences could be arranged for "up" days; unimportant details would be relegated to "down" days. The value of body charting for discovering important biorhythms should not be underestimated. In a January 10, 1972, *Time* magazine article (p. 48) it is reported that "The Swiss have devised a pocket calculator that when individually set will show the owner's 'off' days —when he is accident prone, forgetful or in low spirits. In Japan, the Ohmi Railway Co. has stored in a computer the biorhythms of each of its 500 bus drivers. At the beginning of each shift, drivers scheduled to have 'bad' days are given a card reminding them to be extra careful. In their first biorhythmic year, 1969, Ohmi's drivers achieved a 50% drop in accidents, a downward trend that continued last year."

Although body charting uses only the grossest feedback —experiences, moods, self-observations—it is a very useful adjunct to biofeedback training. With body charting, the individual regards his physiological rhythms as fixed patterns and plans his life around them for the sake of efficiency. With biofeedback training, he actually changes his body state for the sake of mastery. When used in combination, these techniques offer both the option for increased self-control and a method for determining when this control can be used to best advantage. Neither should be regarded as exclusive of the other. To guide your life solely by body rhythms is to surrender a valuable power for autonomy and self-control. To attempt, on the other hand, complete voluntary control of your behavior without some consideration of these rhythms is to accept an enormous and unnecessary burden. Genuine freedom lies somewhere in between.

Fortunately, body charting does not require elaborate machinery. The technique is simple, practical and available to anyone wishing to use it. As of this writing, one company is already advertising body charts for sale. Other firms are certain to provide similar services within the coming months. One of the reasons for this new interest in body charting can be traced to Gay Luce's book, *Body Time,* the most complete and important work on the topic of body rhythms to date. The authors gratefully acknowledge her assistance in constructing the "Body Rhythm Diary."

Your Own Body Rhythm Diary

There are many ways to chart your body cycles, but we suggest the following—it is inexpensive, convenient and can be tailored to meet your own individual needs—a body rhythm master chart, a list of twenty-nine questions about your daily activities and experiences. By duplicating this chart for as many days as you plan to chart your body cycles, you can construct your personalized diary

100

(number the charts consecutively, and use a new one each day).

How many days should you chart your body cycles? The longer the better. As we have already indicated, some cycles span months and even years. A good sampling should embrace a period of ninety days, or three months. The minimum acceptable time limit is thirty days, or one month. As you fill out your daily chart *do not look back on the answers from previous days until you have completed all the charts and are ready to tally the results.* If you do, you may unconsciously bias your answers to fit ideas of what your rhythms "should" be.

Some of the items on the list ask you to make observations at different times of the day. This does not mean that you must constantly be making notes about yourself, though such diligence *would* improve accuracy. Familiarize yourself with the questions, try to become more aware of your daily experiences and then use the chart as follows:

Each page of the diary should cover a twenty-four-hour period, preferably from the time before you usually go to sleep to the same time the next day. The breaking point between twenty-four-hour periods is optional, but, once it is determined, stick to it. *SLEEP AND NAPPING* questions are best answered after rising. Starred questions cannot be answered until the end of the twenty-four-hour cycle. Other questions are best answered as they arise, but if this is impractical they can be answered the same time as the starred questions. Items 17a, 17b and 17c are specifically designed to meet your individual needs. If you have any chronic symptoms, list them here, one after each question. (If you have more than three, add your individual symptoms in the same format—17d, 17e, etc.) These symptoms would include such things as recurrent anxiety, headaches, cramps, eye trouble, stomach aches, fever, asthma, toothache, acne, sinus trouble, allergy, backache, dizziness, cold sores. *Body charting should not preclude a medical examination if serious symptoms persist.* In answering question 23, do not be surprised if you don't experience all the various moods on a given day! Respond only to those items that are relevant for each twenty-four-hour period.

There are many ways completed charts can be used to improve your daily life. The following guidelines are examples:

If you take a long time to go to sleep(2), sleep poorly (3) and require an alarm to get up (5), you may not be sleeping according to your natural body cycle. A change of job hours or routine might be required. Do you constantly feel alert when you get up (6) or do you build up to a state of alertness later in the day (23)? The answers to these questions will help determine when you should arrange business and social engagements that require concentration.

See if your eating habits vary in a consistent pattern from day to day (12, 13). In other words, do you become a glutton every fifth day? This information could be particularly useful to dieters. Instead of trying to eat the same number of calories each day, you could schedule a variable diet that parallels your body's rhythms and therefore would be less likely to break down on days when more food is demanded by your system. With the answers to questions 8 through 11, you could also predict daily hunger pangs and the times when food is more appealing because your taste buds are most sensitive. You could then arrange to be engaged in distracting activities during those periods.

Once you know the daily rhythm of your normal bowel movements (15, 16), you might have the ability to predict the arrival of colds and viruses, which some investigators think are usually preceded by shifts in the time of bowel movements. You can also learn which symptoms (17a, 17b, 17c) have predictable rhythms. This knowledge will allow you to prepare for a recurrence of these symptoms. Asthmatics, for example, could schedule their activities to avoid polluted city areas on certain extra-sensitive days.

Mood charting can be particularly useful. "[With] practice," Luce writes, "you will probably discover that your moods change—not just in response to the outside world,

102

but in weekly, monthly, seasonal, or even six-month cycles. . . . So called moody people . . . can make their lives far smoother by predicting irritation or depression [questions 23 and 29] and consciously planning to compensate rather than feel like a victim of the outer world, and victimizing those around them." [20] Question 23 may reveal that you tend toward certain moods at specific times of the day as well as on specific days of the month or year.

Your daydreams (26, 27) may also follow a regular pattern, just as night dreams come at predictable intervals during sleep. Again, this information about mood shifts and daydream rhythms would allow you to plan your day to best advantage.

At first, the prospect of body charting may seem like a chore. In practice, however, it can become an intriguing adventure. Most of us are so out of touch with ourselves that any extensive attempt at self-observation (*not* self-analysis) produces many unexpected experiences and insights. If you undertake a diligent and extensive program of body charting, two results are certain. First, you will have a powerful tool for improving the quality of your life. Second, your behavior will never be quite the same again.

Not only is charting an effective method for pinpointing body cycles, it can also tell you what behaviors are *not* cycles. If you find that a certain symptom—depression, for example—does not appear in a systematic hourly, daily or monthly pattern, you can begin to search for other causes. Many symptoms are psychosomatic responses to stress that have nothing to do with body rhythms.

In the course of body charting, you may find that your memory is fuzzy about events that took place less than twenty-four hours ago. Instead of being discouraged, use this opportunity to learn more about yourself. Note these blank spots with question marks. You may find that you are consistently unaware of certain events. If so, what are they? Continuous lack of awareness usually means that something about your daily life should be changed. For example, if you can't remember whether you had sexual

103

BODY RHYTHM DIARY: MASTER CHART

Chart #_____

Time: For the 24-hour period   from _____ to _____

SLEEP AND NAPPING

(1)   When did you go to bed? ⎯⎯
(2)   How long did it take you to fall asleep? ⎯⎯
(3)   Did you sleep well? ⎯⎯
(4)   When did you wake up? ⎯⎯
(5)   Did you use an alarm? ⎯⎯
(6)   Did you feel alert? ⎯⎯
(7)   Did you take any naps? If so, when ⎯⎯ ⎯⎯ ⎯⎯ ⎯⎯

APPETITE AND EATING

(8)   At what times did you experience hunger pangs? ⎯⎯ ⎯⎯ ⎯⎯
(9)   When did you have breakfast? ⎯⎯
      Lunch? ⎯⎯
      Dinner? ⎯⎯
(10) * Which meal did you like the best? ⎯⎯
(11) When did you snack? ⎯⎯ ⎯⎯ ⎯⎯
(12) * Total number of snacks? ⎯⎯
(13) * How did you eat during the last 24 hours (check one): ⎯⎯ No appetite
      ⎯⎯ As usual
      ⎯⎯ Glutton

HEALTH (for 17a, 17b and 17c, see text)

(14) Weight before breakfast? (Try to weigh yourself each day) ⎯⎯
(15) Time of bowel movements? ⎯⎯ ⎯⎯ ⎯⎯ ⎯⎯
(16) Nature of bowel movements? (Normal, constipated, diarrhea) ⎯⎯ ⎯⎯ ⎯⎯
(17a) Did you have this symptom today: ⎯⎯
      When? ⎯⎯ ⎯⎯ ⎯⎯ ⎯⎯
(17b) Did you have this symptom today: ⎯⎯
      When? ⎯⎯ ⎯⎯ ⎯⎯ ⎯⎯
(17c) Did you have this symptom today: ⎯⎯
      When? ⎯⎯ ⎯⎯ ⎯⎯ ⎯⎯

ACTIVITY

(18) Type of sexual activity (e.g.,
none, masturbation,
intercourse) during this
24-hour period?   _____ _____ _____ _____

(19)   When?   _____ _____ _____ _____

(20) Type of physical exercise
during this 24-hour period?
(Golf, jog, etc.)   _____ _____ _____ _____

(21)   When?   _____ _____ _____ _____

(22) * On the whole, how would   _____ Clumsy
you describe yourself during   _____ Normal
the last 24 hours?   _____ Adroit

MOODS

(23) At what time did you feel:

alert?	_____	_____	_____	_____
tired?	_____	_____	_____	_____
productive?	_____	_____	_____	_____
mentally dull?	_____	_____	_____	_____
athletic?	_____	_____	_____	_____
agitated?	_____	_____	_____	_____
happy?	_____	_____	_____	_____
depressed?	_____	_____	_____	_____
sensitive?	_____	_____	_____	_____
disoriented?	_____	_____	_____	_____

(24) At what times did you
feel like a smoke?   _____ _____ _____ _____

(25) Did you have any good
fantasies?   When?   _____ _____ _____ _____

(26) Did you have any bad
fantasies?   When?   _____ _____ _____ _____

(27) * On the whole, could you
concentrate easily   _____ _____ _____ _____
were you easily distracted _____ _____ _____ _____
or normal   _____ _____ _____ _____
during the last 24 hours?

(28) * Were you more introverted _____ _____ _____ _____
or extroverted   _____ _____ _____ _____
during the last 24 hours?

(29) * How would you summarize
your mood during the last
24 hours? (Happy, sad,
joyous, despondent,
listless, etc.)   _____

105

activity within the last day, your relationship with your mate may be in trouble.

Different Kinds of Feedback

There is growing evidence that many kinds of feedback, not simply *biological* feedback, have a healthy, self-correcting influence on an individual's behavior. These include visual, auditory, pencil-and-paper and interpersonal feedback. The visual form of feedback, involving video-tape equipment (a do-it-yourself TV production kit), is not readily available to the average consumer but soon will be, judging from recent developments and price cuts in videotape units. We include it for this reason. The other types of feedback we will be discussing can be used by the average American right now.

Visual Feedback

We've all met one: the perfect brat. For him, temper tantrums are a way of life. He demands everything, usually in nerve-piercing tones, and gives you nothing in return except tension headaches. The targets of his frenzied assaults are limitless: the neighbor's cat, your new dishes, even his teacher's fragile shins. "How can his mother tolerate this disaster?" her neighbors ask. "If she could only see how other people view her meek responses to the little monster!"

Jeff's mother had the chance.[21] After eight and a half years of coping with a child whose attitude toward life was "I have the right to do anything I want to" she finally went to psychologist Martha Bernal's office for consultation. A program was worked out that combined conditioning methods with televised feedback of her relationship with Jeffrey. Several videotapes of the mother and Jeff interacting at home and at the psychologist's office were produced and then reviewed by the mother and Dr. Bernal in private meetings.

106

The power of television feedback was revealed in the very first screening of "Life with Jeff." After watching only thirty minutes of her pitifully meek responses to Jeff's incessant demands, she turned to the clinician and exclaimed of herself, "What a dishrag!" "Never again, in all subsequent observations of her and Jeff," recalls Dr. Bernal, "did she ever show the same 'dishrag' quality; gradually she became more assertive with Jeff as well as with other people." [22] Using additional videotape feedback along with instructions from Dr. Bernal, the mother was finally able to bring the brat under control. After eighteen weeks and only seven TV sessions, Jeff could be considered at least as well behaved as the average child his age. "Two years later," reports Dr. Bernal, "the mother-son relationship was vastly different. Jeff was courteous and affectionate, and the mother felt a genuine affection for him." [23]

The change in Jeff's mother is not unique. Videotape is a feedback tool that is becoming increasingly popular, particularly in therapeutic and educational institutions.

The power of visual feedback seems to be its ability to make a person aware of those actions that cannot be properly expressed in words or which the person simply refuses to "see." As one encounter-group participant aptly stated while pointing angrily at a videotape playbook set, "You can't run away from this thing."

Encounter-group leader Dr. Frederick Stoller describes the case of a woman who had spent a significant part of one group session lamenting that her husband "behaved like a child" in her presence. "On videotape, I was able to show her that she used many of the mannerisms of a scolding mother with him," writes Dr. Stoller. "She would glare, shake her finger, and, when pleased, pat his head." [24] The woman's amazed response: "I couldn't believe it. It was worth a thousand words."

Sports fans, of course, will not be surprised with statements endorsing the effectiveness of videotape feedback. Professional football coaches have used filmed feedback for years to help players improve their performances on the field. Similarly, baseball and golf pros employ instant-replay videotape equipment to perfect their horizontal and vertical strokes.

Visual feedback need not be elaborate to be effective. Clinical researchers Floyd Cornelison and John Arsenian use a simple camera. They report that "in some cases, psychotic patients who looked at photographs of themselves underwent dramatic improvements." [25] The use of videotape need not be confined to structured therapeutic or educational settings. Neither should you require a specific reason or goal in order to benefit from using it. Simply using it around the house—taping your family and friends and, in turn, having them tape you—can be an enlightening and helpful activity. In a surprisingly short space of time, the novelty of being "on camera" will wear off. The machinery will become unobtrusive, and you will then have the opportunity to get feedback on your normal everyday behavior. (Think how relaxed Americans have become in front of cameras installed in banks and shopping centers.) The effects of this exposure—to your mannerisms, role-playing behaviors, posture—are hard to describe but are often quite dramatic. You may think that you already know yourself, but so did Jeff's mother.

Auditory Feedback

Although visual monitoring appears to be the most potent form of mechanical feedback, it is not always practical or appropriate. To use videotape at home or in a therapist's office is one thing, but to have a camera crew follow you through your daily activities would be both expensive and inconvenient. A more flexible feedback device is the audiotape recorder, which can be carried almost anywhere (as teen-agers prove every day) and is so well integrated into our pop culture that most people do not feel out of place using it.

Although many people feel uncomfortable, even anxious, when confronted with their own voice, auditory feedback is ultimately beneficial in a variety of contexts including the home, the classroom and the job. For example, one scientist has shown that people learn new material much sooner when it is presented with their own voice rather than someone else's.[26] Listening to differences in your voice as you talk to important people in your life, such as parents, business subordinates, boss, wife (or husband) and children, is often a stunning educational experience. Also, you'd be surprised to see how easily disputed points can be cleared up when an argument is on tape.

Pencil-and-Paper Feedback

Sometimes feedback can be accomplished without machines at all. Dr. Richard McFall of the University of Wisconsin made the unexpected discovery that even brief but systematic self-observations without mechanical equipment can be effective feedback. As part of a large experiment on smoking he had instructed a group of volunteers to keep hour-by-hour records of their normal smoking behavior in small booklets. Several of the subjects claimed that, as a result of self-monitoring alone, they smoked progressively fewer cigarettes with each succeeding day. "This occurred," notes McFall, "despite specific instructions that they were not to alter their smoking behavior in any way during this period."[27] This finding has some interesting implications for body charting. If you are overweight and begin to faithfully record everything you eat, you might eat less.

Interpersonal Feedback

Finally, we come to the most common feedback mechanism known to man: man himself. Almost all the feedback we ever get—information about how we look, how well we perform different tasks, what we are unconsciously doing—comes from the people around us: parents, friends, children, marriage partners and colleagues. Unfortunately, people are not perfect mirrors of their fellow men. They have a tendency to withhold and to distort information—even lie—sometimes with disastrous results, as in the case of the helpless child whose self-image is created by two neurotic parents. On those rare occasions, however, when people do give accurate feedback (that is, when they say what they see) the results can be spectacular.

Every experienced therapist, regardless of his theoretical persuasion, knows that reflecting what he sees in a client's behavior, honestly and without judgment, can have potent curative effects. Chris Wendell, a graduate student at the California School of Professional Psychology, demonstrated this recently in a simple experiment. Under the supervision of a licensed psychologist, she joined a sensitivity group with the intention of observing and reporting the physical movements of each member. Just by making direct, uncomplicated statements, such as "I see you hunching your body," "The left part of your body seems frozen," "You seem to be holding yourself in," Ms. Wendell was able to help people unblock a wide range of physical and emotional rigidities.[28]

Many years ago, a prominent clinician named Carl Rogers developed an entire therapeutic system based on the idea of direct interpersonal feedback. A client would state his problems, feelings and emotions. Rogers would just repeat or rephrase what was said, sometimes adding his perceptions of the client's nonverbal responses (body language). Rogers believed that a deep Freudian analysis to dredge up unconscious material or a complex conditioning program might be useful, but he put his primary trust in the ability of the human organism, given accurate feedback, to regulate itself in its own best interest. Rogers's Client-Centered Therapy is now taught at psychology departments and medical schools across the country and has clearly influenced other forms of psychotherapy as well.

Because feedback from different persons is so important to healthy functioning, Dr. Stoller believes that the small family unit is obsolete. "The current American family is becoming increasingly unique," he writes, "in that it sees itself as an independent unit, operating without reliance on the extended family—grandparents, uncles, aunts, cousins. This is an impossible task."[29] He predicts the evolution of "colleges for growth" where "deprived"

110

children from affluent nuclear families can get the diversified interpersonal feedback they so desperately need.

As we have shown, there is mounting evidence that feedback, not just biological feedback, has broad corrective powers. When given nonjudgmental feedback, whether in the form of bleeps on an EEG machine or straightforward comments from a friend, the average person seems to gain greater control over his own behavior and, furthermore, utilizes this control in his own best interest.

Of course, this finding goes against covert but strongly enforced cultural assumptions. Contrary to the lip service given such ideals as honesty, close personal contact, and letting people "do their own thing," our society does not have faith in man's inherent ability to manage his own destiny, given the information he needs to attain self-control and self-sufficiency. Americans spend billions of dollars each year through the educational system alone to make sure that the nation's children are properly instilled with the correct values and morality. Despite the obvious failure of this system, one popular psychologist perversely has argued that cultural stability requires even more subtle and efficient conditioning of the young. [30]

We do not mean to imply that children should be given EMG machines and a mirror and then left to fend for themselves. We are simply suggesting that the results of feedback experiments give us good cause to reevaluate some basic assumptions, particularly in the area of child rearing and interpersonal relations. Do we really help someone by telling polite lies? Do we really help children by trying to force them into some ideal mold, or is it better to accept and reflect these youngsters as they are?

More importantly, feedback research should cause us to reconsider the relative worth of what we have to give to others. Liberal education has taught us to be witty, insightful and incisive in order to be "good" company and "valuable" friends. When someone has a problem, we immediately offer our judgments, our morals, our philosophies, our complex prescriptions. However, it may be that the most important gift we have to give is, quite simply, what we see.

111

Underground Science

Less than one hundred years ago an American news-paper printed the following editorial:

"A man was arrested yesterday, charged with at-tempting to obtain money under false pretenses. He claimed he was promoting a device whereby one person could talk to another several miles away, by means of a small apparatus and some wire. Without doubt this man is a fraud and an unscrupulous trickster and must be taught that the American public is too smart to be the victim of this and similar schemes. Even if this insane idea worked it would have no practical value other than for circus sideshows."

A short time later, in 1876, Alexander Graham Bell took out a patent on the telephone.[1]

What You Won't Read in the Journals

It is obvious to us that the implications of bio-feedback research go far beyond the conservative specula-tions of staid scientific journals. Many times during pri-vate meetings with biofeedback researchers, we were inevitably led from discussions of voluntary control to such topics as extrasensory perception, psychic healing and psychokinesis. Publicly, scientists are doing conventional research on the applications of biofeedback training; privately, many acknowledge more unorthodox interests.

Why the secrecy? The primary reason is the fear of subtle reprisal from the academic establishment. Many unconventional explorers are affiliated with universities where employment and success depend upon recognition from (read: "acceptance by") one's peers. Although most scientists pride themselves on being open-minded, underground scientists are aware that the recent history of scientific innovations tells another story. Rebels against established science are not burned at the stake or humiliated as severely as Galileo was, but the tortures of professional scorn are grueling nevertheless. "Radical innovators can still be ostracized, prevented from speaking about or publishing their ideas, cut off from research money, fired, even jailed." [2]

For the remainder of this chapter we will discuss the relationship between biofeedback research and some of the more unorthodox areas of science. If you should be skeptical of the following material, please understand that the authors are sympathetic. We, too, have had our qualms and are frequently torn between the facts of this unusual research and our own academically bred conservatism. We do believe, however, that at this point in history it is irrelevant to debate over which aspects of underground science are real and which are not. The data are simply not conclusive enough to arrive at final decisions. Two things are important. First, scientists and laymen should be aware that some progress has been made in parapsychology and other underground areas. Second, further research should be conducted to sharpen our understanding of this new realm of knowledge.

Three from the Underground

By day Russell Targ is a straight scientist, and a good one at that. From nine to five he conducts high-level laser research at Sylvania Laboratories in Mountain View, California. But at night, he enters a different world of science: the world of parapsychology. Russell Targ, along with his associate David Hart, has invented a machine that teaches people ESP ability. The machine, which operates on the simple principle of giving a person immediate feedback (knowledge of results) for each ESP attempt, has significantly improved ESP ability in over 30 percent of the people tested.

Just a few blocks away from Targ's home is the Stanford University office of Dr. William A. Tiller. Dr. Tiller is a full professor of materials science. His academic honors include a doctorate in physics from the University of Toronto, the chairmanship of the Materials Science Department at Stanford and a Guggenheim fellowship. His present research, however, is motivated by an unusual apprehension: "I must confess that one of the greatest concerns I have is that the Russians are tapping [psychic] powers that are so far beyond anything that we know of and that the possibility exists that these powers may provide a way for world domination that has not yet been considered." Dr. Tiller believes not only that psychic powers are real but that they are genetically inherited, a factor which favors the Russians. "In the West," he explains, "we did away with a lot of our witches, we had a lot of witch hunts and a lot of inquisitions, and so on. And so I wonder about our gene bank. . . . The Eastern part of Europe, you see, didn't have that. So that's one of the ways that the Russians are better off. [8] Dr. Tiller spent a good part of last year reviewing Russian work in parapsychology. He is now attempting to duplicate some of their machinery and experiments here in the United States.

From Stanford it is a short drive along the rim of the Los Altos Hills to San Jose and the home of Marcel J. Vogel. To most of his colleagues at the nearby I.B.M.

research center, Marcel is an expert in the field of solid state physics. (He is the author of two internationally known books on energy conversion in crystalline solids.) But to those who know of his underground science interests, Marcel Vogel is the Plant Man. Marcel talks with plants. Furthermore, he has constructed a device which allows him to pick up the physiological reactions of plants to human behavior. "Plants respond to strong emotions," he claims.[4] These include both "good vibrations" such as love and "bad vibes" such as hate and aggression.

Russell Targ, William Tiller and Marcel Vogel are three members of a growing scientific underground in the United States. They are, according to Stanford research physicist Dr. Hal Putoff, some of the "many people in this country who sort of run a double life in their scientific work."[5] Dr. Putoff should know. He divides his time between "legitimate" research and investigations into such areas as ESP, body auras and scientology.

Despite the opposition of establishment science to many aspects of underground science the line between the two is beginning to fade. There are several reasons for this development. For one, the instances of "impossible" events are becoming hard to ignore. The evidence for ESP alone is much more ironclad than for some of the more commonly accepted psychological phenomena.[6] A second factor is the impact of the Russian research in parapsychology. Americans first became aware of the Russian work in 1970 when Sheila Ostrander and Lynn Schroeder published *Psychic Discoveries Behind the Iron Curtain*. Since then, numerous American scientists have traveled to Russia to check out the validity of the information reported in the book. These "second wave" reporters disagree about the ultimate political consequences of the Russian experiments, but there is unanimous agreement on one point: the Soviet scientists have made substantial progress in both the areas of ESP and psychokinesis (using psychic powers to move objects).

The key factor, however, in making underground science legitimate, at least to a larger number of scientists and laymen if not to the majority of the scientific establishment, has been biofeedback research. Biofeedback training has demonstrated that a lot of activity which traditional psychology has denied, or at best ignored—behaviors such as shifting into altered states of consciousness, controlling the internal organs, changing brain waves and body radiations at will—are real. Once we understand that man can turn his alpha waves on and off just as easily as if he were pressing a telegraph key, it is no longer as difficult to imagine that some brain waves might be patterned into message form and then recorded by another sensitive brain. More importantly, biofeedback research has developed a technology that can monitor mental energies and translate them into clicks, bleeps and bips on precision instruments. In short, there is now a machinery to translate those weird, subjective and nutty parapsychological ideas into something real, something that can be analyzed, interpreted, studied, duplicated and, of course, published in the scientific journals.

There is an interesting analogy between the impact of biofeedback research on parapsychology and the first developments in dream research. Less than fifteen years ago, dreams were regarded as fluke occurrences by almost all theoretical scientists and doctors. Only the Freudian fringe and a few other psychotherapists talked about the importance of dreams. Then some investigators [7] observed that people's subjective estimates of their dream length correlated with bursts of eye movements that have come to be known as rapid eye movement (REM) sleep. All of a sudden dreams, instead of being those funny subjective experiences, became real. Dreams could be measured, so they were valid objects for study. This new legitimacy has led to further physiological research which has been very valuable. It has also increased purely psychological research into dreams, research that could have been conducted thirty years earlier but was not supported because it was not quite respectable,

ESP or SSP?

There is a long history of underground research. Unfortunately, psychic phenomena in the West (particularly in America) have always been associated with religious or spiritual ideas. This distinction [8] is inherent in the Judeo-Christian tradition, which divides the study of man into two distinct parts: the investigation of bodily mechanics and the study of the interaction of mind and soul, or theology. The positive consequences of dividing man into body and spirit are well known to historians of science. When Descartes pinned the soul to the pineal gland at the front of the brain (c. 1650), he freed scientists to explore the machinery of the human animal without violating ecclesiastical precepts. However, the negative consequences of such a distinction have never been fully appreciated. During the centuries that preceded Descartes, all mental phenomena—including psychic phenomena—became inextricably confused with the notion of soul. To deny one automatically implied a denial of the other. The result is that, as scientists eliminated the need for soul as an explanatory concept in psychology, the belief in psychic events was simultaneously eliminated. Instead of extending science to account for ESP and other inexplicable mental events, investigators dismissed them as either nonexistent or illusory. Even worse, those researchers who continued to believe in psychic phenomena bought the idea that paranormal events must exist outside the realm of science.

For reasons that are not entirely clear, the Russians do not make a distinction between science, on the one hand, and psychic phenomena, on the other. The Russian scientists take the primary view that there must be a logical explanation for everything. If, for example, a Soviet scientist hears that some crazy student who has locked himself in a concrete vault is able to "visualize" events outside the vault, the scientist will immediately begin to search for a physical theory to explain what has happened.

Technically, then, the Russians do not believe in *extrasensory perception,* which implies a method of communication outside the scope of scientific explanation. Rather, they subscribe to the idea of *supersensory perception* (SSP), a way of communicating which is not yet understood but which is, in theory, explainable in scientific terms. This assumption—that psychic phenomena have a physical basis—has led many Soviet scientists to investigate the kinds of strange events that their American counterparts would frequently ignore.

In this context, the chief contribution of biofeedback research to parapsychology in the West is to provide a physicalistic model of ESP. We now know that a person produces electromagnetic body waves that can, with practice, be controlled in such a way as to provide a coded message.[9] We also know that these waves can be transmitted and received at a distant point. For example, Dr. W. A. Schafer, formerly of General Dynamics Life Sciences, has developed a complex of supersensitive electronic instruments with the descriptive but unimaginative name of Field Effect Monitors. Using these devices (which are now on the commercial market) he can pick up electromagnetic waves produced by heartbeats at over four feet from the body.[10] When amplifiers are added to the circuit, the distance that body signals can travel is greatly increased.

As a result of biofeedback research, three body-wave theories of ESP have received serious attention. The most popular, and least likely, is the theory that ESP is transmitted by alpha waves. Why alpha? Of all the brain states, alpha is the easiest to bring under voluntary control. Furthermore, there is some experimental evidence linking alpha to ESP. A few years ago two Philadelphia researchers made some interesting observations. They were studying the brain-wave patterns of two identical twins who were situated in different rooms. When one twin closed his eyes, the EEG machine registered a common alpha response. But something uncommon also happened. Whenever the first twin closed his eyes, the second twin experienced alpha, although his eyes remained open.[11] Despite this evidence, however, the case against alpha as the *sole* cause of ESP is strong. "It must be stressed," writes Joe Kamiya, "that there is no connection between alpha waves and extrasensory perception. People tend to associate the two because radio waves are involved in communication, but radio waves are generated at several thousand cycles per second, while brain waves range between a fraction of a cycle and about 100 cycles per second, with most of the energy limited to about 15 cycles per second."[12] The alpha signal is so weak relative to other energies it is hard to believe that alpha, or at least alpha alone, is the energy source for psychic phenomena.

A more promising variation on the alpha theory holds that ESP is a product of different body rhythms operating in some kind of synchronous pattern. A good way to understand this idea is to consider the following example. Take two metals. Both of them have a particular melting point. If you combine these metals, however, the result is a much different melting point than either of the two had previously.[13] This process of combining a number of different things to achieve a "higher order" result is called a "synergistic effect." Similarly, the synergistic theory of ESP holds that psychic communication is really a higher-order effect resulting from the combination of normal body energies, such as brain waves, heart rate and galvanic skin response, into a synchronous pattern. Dr. Tiller believes that biofeedback training will be useful for helping us to synchronize our glands so that psychic phenomena can be controlled systematically and at will.[14]

The third body-wave theory of ESP is the most exotic. It is also the most probable, given what we already know from conventional science. It developed a few years ago when physicist Gerald Feinberg became bored with the idea that Einstein's light speed is the maximum allowed for our universe. Employing some clever mathematical techniques, Feinberg invented a proof for the existence of a faster-than-light particle, which he calls the "tachyon." Many scientists now believe not only that the tachyon exists but that it is the physical basis of ESP. At least one prominent physicist currently investigating tachyons as part of an elaborate government research project is also doing underground research on the relationship between tachyon waves and ESP.

Beyond theoretical considerations, biofeedback training has actually proved useful in helping people to gain voluntary control over their extrasensory perception abilities. Here we get into a more detailed description of Russell Targ's ESP teaching machine. The principle behind the machine is quite simple. For several years, ESP researchers have observed what they call the "decline effect," the phenomenon where a subject in an ESP experiment will start off demonstrating a high clairvoyant ability but gradually will lose that ability. Targ hypothesized that this shift might be due to boredom, since in most ESP experiments a subject is not given knowledge of his results until the experiment is over. For example, if you were asked to guess which cards were being turned up as someone else in another room was going through a deck of them, you would not receive your "score" until all fifty-two cards had been dealt. Targ reasoned that ESP ability could be improved if a subject could get immediate feedback after each ESP attempt. In Targ's words, "The immediate reinforcement gives the subject a feel for what the ESP is like." [15]

Targ's machine looks like a small toolbox with four 35-millimeter transparencies and four corresponding buttons built into the face. When the machine is turned on, it automatically and randomly selects one of the four transparencies. The subject tries to guess what choice the machine has made and then presses one of the four buttons. The correct transparency lights up, and the subject knows immediately if his guess was right or wrong. (By pressing a "pass" button the subject may also choose to skip a trial if he wishes.) After twenty-four trials, the machine places a successful subject's ESP ability in one of four categories with "encouragement lights": ESP ability present," "useful at Vegas," "outstanding ESP ability" and "psychic medium oracle."

As we indicated earlier, repeated use of the machine increased extrasensory perception in about 30 percent of the people tested. Targ concludes from this that "it is possible to teach and enhance ESP phenomena through techniques of feedback and reward in much the same way as visceral and glandular functions are brought under volitional control." [16] He is now undertaking experiments to find the relationship between the accuracy of precognition (foretelling future events) and the temporal distance from the event.

Psychokinesis: Mind over Matter

The place is a scientific laboratory in the Soviet Union. The scene is as follows: a female subject is asked to move a small cigar box cylinder a short distance. She does as she is instructed—without touching the object in any way. The woman has demonstrated psychokinesis (PK), the ability to move objects through psychic powers. She is not the only person possessing such talent. Dr. Gertrude Schmeidler, Professor of Psychology at the City University of New York, reports an experiment where a subject ". . . used PK to make a pointer shielded from air currents track curves on smoked paper." [17] As research evidence accumulates, it is becoming increasingly evident that man can use the power of his mind to manipulate physical objects—an ability once thought to exist only in the realm of science fiction.

Thus far the impact of biofeedback research on PK research in the United States has been indirect. There are no feedback techniques that we know of which can improve PK ability. What has happened, however, is that investigations into the relationships of biofeedback training and meditation are bringing people with PK ability into the laboratory for the first time. Scientists originally sought Yogis, swamis and other meditators to measure their physiological responses on EEG and EMG machines. These investigators, many of whom would never think of studying PK ability, are suddenly astounded when these

121

meditators begin proving their claims that they can move objects around the laboratory. As a result of these surprises, a small group of American scientists are now undertaking systematic research into psychokinesis.

As in the case of ESP, American research lags far behind the Soviet effort when it comes to investigating PK. While in Russia, Sheila Ostrander and Lynn Schroeder jotted down the following notes on an experimental study of Nelya Mikhailova, a woman with highly reputed PK abilities:

> Prior to a demonstration of PK effects she is thoroughly examined by a doctor and double checked by x-ray . . . she affects a compass needle inside of a plastic case on a leather strap . . . sometimes takes two to four hours to rev up enough energy to make the needle turn . . . during filmed demonstration it took about twenty minutes . . . intense strain . . . heartbeat to 250 beats/minute . . . compass needle begins to turn, then the entire compass turns, usually counterclockwise and spins like a top. Mikhailova can move various non-magnetic objects outside of or inside of a transparent plastic box.[18]

While visiting Russia to check the findings of Ostrander and Schroeder, Dr. William Tiller observed a film of Alla Vinogradov, a trained PK performer, who is able to move objects without any sign of stress. Vinogradov can provoke rolling movements in round objects of up to 200 grams in weight. She can also drag objects with PK but finds this a greater strain. Describing her subjective feelings during PK experiments, Vinogradov claims that she does not usually feel stress. She believes the energy comes from her solar plexus and feels that anyone can train himself to do PK. When trying to roll objects diagonally, she experiences an amplification of energy across her forehead. When she senses that skeptics are present, she feels she must expend more energy to do the best she can.[19]

Dr. Tiller is convinced that the Russians are undertaking a full-scale investigation into the physics of PK. He has discovered that a new Institute for Psychology will be opened soon under the auspices of the Soviet Academy of Sciences and that the institute will focus on studying the controllability of PK phenomena.

The Electromagnetic Personality

The more we study the rhythms of nature, the more we realize that man is not a self-contained unit. Man is not just a clock within himself, he is part of a much larger timepiece called the universe. His abilities, moods and even thoughts are influenced by the ebb and flow of celestial mechanics: the slow spinning of the earth around its axis, the elliptical swing of the moon, the pulsating flairs and electromagnetic energy from the sun. Our ancestors knew intuitively that man was a reflection of his environment. The ancient Greeks defined physical health as being in harmony with nature. Many Eastern religions advised followers to sleep with their heads to the North and their feet to the South; in this way, the body could gain strength from the earth's magnetic currents flowing in their "proper" direction. Astrology, of course, is based on the assumption that human behavior echoes the rhythm of heavenly orbs.

The secular religion of science is just beginning to document this atavistic wisdom. A few observers have noticed that more people are admitted to mental institutions, and more crimes of violence are committed, during the full moon. Ms. Margaret Harrell of the Alameda County Blood Bank in Oakland, California, supported this observation when she recently told viewers of a local TV program that doctors get more crank telephone calls during the full moon than at any other time.[20] "These callers seem to be affected by the full moon," she reported. "A woman phones a San Francisco hospital saying that the L.A. Police Department is causing her to bleed. Others claim that nurses are stealing their furniture. Some even try to give away money. It starts about five days before the full moon and ends about five days after." In her fascinating

book *Body Time,* Gay Luce describes the impact of light/
dark cycles on human performance. Among other things,
light activates the adrenal glands, causes radical changes
in brain chemistry and even alters the size of sexual
organs in male animals.[21] When the pattern of the light/
dark cycle is disturbed, as after a long jet flight across
time zones, man's body rhythms are thrown drastically out
of "sync."

Most scientists now accept the fact that human behavior is influenced by geophysical rhythms. The question
is, to what extent? The results of biofeedback research
indicate that man is far more sensitive to subtle envir-
onmental changes than most scientists have let themselves
believe. It is interesting to note, for example, that there is
a strong correlation between brain-wave states and variations in the earth's electromagnetic field. In other words,
the earth pulsates different electromagnetic energies of
specific intensities. The frequencies of these pulsations
correspond directly with our own alpha, beta, delta and
theta waves.[22] One explanation for this correspondence is
that tuning in to the earth's vibrations is a necessary part
of biological evolution. This would explain why animals
have built-in alarm systems that allow them to survive
earthquakes. An earthquake causes drastic changes in the
earth's magnetic and electrical fields. This field change
moves at the speed of light, while the quake itself travels
at about the speed of sound, or slower. Well-adapted animals sense a change in the earth's electromagnetic field
before the quake hits and are thereby prepared for
danger.[23]

Just as the large geophysical fields of the earth, moon
and sun influence man's behavior, so do localized electromagnetic fields, such as those created by weather conditions and certain electrical machines. Scientists know, for
example, that boredom and restlessness can be controlled
by changing the electrostatic field where a person is living.
(This can be done quite simply with commercially available equipment.) The positive electrostatic field (high in
negative ions) produces an electrical current in the body,
exciting the nervous system, increasing the nerve impulse

rate to the center of the brain and thus making a person more alert and aware. NASA's Jim Beal notes that in Israel, where civil order is frequently disrupted by hot, dry winds, which are high in positive ions (a negative electrostatic field), the authorities are installing negative ion generators in public buildings.[24]

Researchers disagree, however, about the overall impact of localized electromagnetic fields. An underground minority believe that the entire urban environment, which concentrates so much metal and electrical voltage into a small space, is one large electric depressant. They also believe that a lot of supposedly neurotic problems are the result of sensitivity to electromagnetic fields. "There is a case on record," reports Jim Beal, "of a woman who was picking up noise around her refrigerator, around the electric lights, and around the telephone just before it rang. . . . To solve the problem they finally had to shield everything in the house." [25] This meant they had to put metal conduits around all the electrical wires and appliances.

Then there are people in mental hospitals who claim they cannot get away from noise. They stuff their ears with cotton, but it does not help. Quite often, as they wander around the hospital ground they find a place that is suddenly quiet. Conceivably this could be an electrical null point; in other words, a place where electrical fields are canceling each other out.[26]

How much of our behavior is influenced by the local electromagnetic fields that are so much a part of our lives? Do we set the electric clocks, or do they set us? The answer is likely to come from biofeedback research for the following reason: the electrical frequencies in the atmosphere that influence behavior are strikingly similar to the frequencies which our own brains are capable of producing. Using feedback we may be able to train people to reproduce under experimental conditions the same wave patterns caused by changing electromagnetic fields. We can then observe how each person is affected as he changes his brain waves to duplicate the effects of different external radiations.[27] We may even be able to reproduce the pleasant and healthful effects of positive electrical fields by tapping our own internal electrostatic generator.

Just as in ESP and PK research, the Russians are doing quite a bit of advanced work on the effects of electromagnetic fields. Much of this research is hidden under such titles as "meteorological feeling," "bioenergetic therapy" and "biological radio communication." One area where the Soviets have had particular success is "electrosleep." Essentially this process involves applying a pulsating, low-frequency electrical current to a person's head. The result is a pleasurable experience. Depending on the power used, electrosleep can induce relaxation, sleep or anesthesia. The Soviets have been using this technique in surgery for over ten years, even though they are not quite sure how it works.[28]

An Organic Feedback Machine

As I (Andrews) first entered the home of Marcel Vogel, I was naturally skeptical. When we had arranged our meeting by telephone the day before, he had informed me that I would have an opportunity to communicate with, of all things, a plant. As Marcel showed me into his spacious living room, I was immediately struck by the sight of a large, green, leafy plant which seemed to dwarf even the piano. Attached to the plant were two small electrodes, which in turn were connected to a polygraphlike recorder.

During the first few minutes of my visit, we talked about the results of Vogel's research on plants. As we spoke, a marker on top of the recorder vibrated back and forth across a continuously moving reel of paper. I found Marcel's explanation of how plants respond to electromagnetic radiations fascinating, for the most part. Occasionally, however, I experienced a quiet but forceful internal constriction, a sense of outrage at some of his comments. "Plants respond to human emotions!" Incredulity. "Plants can tell what mood you're in!" Even harder to believe. Suddenly Marcel stopped talking. It was then I realized that the plant *had* read my thoughts! Each time I had experienced a strong doubt the recorder had registered a long red spike.

Until recently, researchers thought biofeedback could only be recorded with sensitive metal and plastic instruments, such as the EEG and EMG, and that recordings could only be made by skin contact. Marcel Vogel has disproved this assumption. Along with a half dozen other investigators, Marcel is showing that plants do more than routinely supply oxygen: they can be "sensitive" companions as well.

Research into the use of plants as long-distance feedback sensors makes fascinating reading. In one experiment, three plants, each attached to recorders, were placed in three different rooms. At another end of the building a specially constructed machine would randomly dump small numbers of tiny shrimp into boiling water. The shrimp died immediately on contact with the water. Precision timing instruments showed that each plant registered an emotional reaction at the exact moment of the shrimp's death. This experiment was replicated several times under rigid scientific controls; the results were always the same.[29] Plants seem to be sensitive to energy waves emanating from living organisms. When these waves change because of biological alterations within the organism, the plant detects this pattern and registers a change.

On the one hand, the use of such "organic biofeedback sensors" is limited. Machines seem to be capable of much finer discriminations. Furthermore, there is the additional problem of translating the plant's reactions into blips, clicks or some other kind of readable data. On the other hand, plants have one distinct advantage: anonymity. For those who want to measure biofeedback without being obvious, the plants present some interesting possibilities. Suppose you are fixed up with a blind date and you want to know what his (or her) first impression of you really is. An obliging plant could tell you if the friendly smile is an unemotionally involved polite gesture or a reflection of genuine interest. Then there is the whole area of political and industrial espionage. A few years from now, a C.I.A.

127

official may not be surprised to learn that the flower on his secretary's desk is really a "plant."

The foregoing, of course, represents fanciful speculation. Nevertheless, biofeedback research with plants has given a boost to underground scientists working on some of the more exotic, less "respectable," areas of plant communication. Some underground scientists, for example, are trying to show that plants have memories. Consider this experiment by Clive Backster. A "plant murderer" was arbitrarily selected from a group of six men. Each of the six, in random order, walked into a room which was inhabited by two plants. When all the visits had been completed only one plant was left alive; the other had been torn from its pot, mangled and finally crushed into the floor by one of the visitors. The only witness to this gruesome event: the remaining philodendron. Spotting the criminal was simple, however. Backster attached electrodes to the plant, then had the six suspects walk into the room in lineup fashion. When the murderer came through the door, the plant emitted the electrical equivalent of a shriek.[30]

Other underground experimenters are trying to demonstrate that plants and humans are capable of emotional relationships, such as love, hate and even religious communion. Marcel Vogel has already found at least six people who have the ability to tune into plant consciousness. "They can mentally go into a plant," reports Vogel. "They and the plant become one. These people can clearly describe the cells, the finer detail of the cells, the color, and even the movement of moisture up the leaf." [31]

Then there are the experiments of Reverend Franklin Loehr of the Religious Research Foundation. With the assistance of over 150 people, Reverend Loehr has conducted more than 700 experiments, using 27,000 seeds and seedlings, on the effects of prayer on plants. Reverend Loehr usually divides helpers into two groups, each given identical seeds and identical soil. One group plants the seeds and prays for their well-being; the other group just plants the seeds. The prayer-nurtured seeds usually grow three or four times faster than the "theologically deprived" seeds.[32]

Marcel Vogel believes that because of biofeedback research both scientists and laymen will be quicker to accept plants as an important teaching tool. "From plants we will learn the subtle power of our own thoughts. [They will teach us] . . . to be simple, to be kind . . . and to stop thinking negatively about people." [33]

Does Man Play Dice with the Universe?

Biofeedback research has challenged a fundamental scientific concept: that all human behavior must inevitably follow laws and patterns. The gloomy picture of man as a machine is given way to a more flattering portrait of man as, if not master, at least junior architect of his own destiny. The staggering implications of this for developments in science, health, education and even politics will be considered in the final chapter. But as long as we're talking about underground science, we would like to go beyond even "staggering" for the moment and offer the following speculation for your consideration and possible amusement.

To say that man is not ruled by the laws of science suggests the possibility that man can, by changing his own behavior, change the laws of science. Gaining control of the nervous system may be another way of saying that you can legislate your own biological laws. In short, if man is not the pawn of fate, perhaps fate is the pawn of man.

Needless to say, this hypothesis raises some fascinating issues. Does man discover the laws of his own biology or does he make them? And what of our culture's preoccupation with the future? Is Alvin Toffler predicting "future shock," or is he writing the script for us to follow? (What is a self-fulfilling prophecy, anyway?) We all know that what was science fiction yesterday is science today and society tomorrow. Is this coincidence or a deliberate attempt to make our dreams come true?

The evidence for this hypothesis is sparse, but it is

intriguing, nevertheless. The best documentation for man's altering the laws of physics is probably Joseph C. Pearce's research on fire walking.[34] Once a year, he reports, Hindus from all over the island of Ceylon gather at a central place to honor the god Kataragama. The ceremony begins in the afternoon when native women walk up and down in front of the temple carrying in their bare hands iron pots filled with burning coconut husks. This is just a prelude, however, to the main event. At night, giant, hardwood logs filling a pit twenty feet long by six feet wide are set afire. The logs burn until they become a deep bed of hot charcoal. Early in the morning, just when the heat is so intense that it is difficult to breathe within a few yards of the pit, drums build to a climax, the huge temple doors swing open, and scores of Hindus, both priests and initiates, stream out of the temple and straight into the pit. The majority survive unscathed; a minority—"a macabre control group"—suffer serious burns and even death.[35]

The studies of fire walking are so numerous that they cannot be ignored or dismissed as illusion. As far back as 1936, the English Society for Psychical Research had two Indian fakirs walk on fire beds with surface temperatures up to 500 degrees C. Under the watchful eyes of doctors, scientists and psychologists from Oxford University —and with no chemical preparations—the Indians walked the fire over and over again for a period of several weeks.[36]

Then there is the case of Arigo, the famous Brazilian healer. Before his death a few years ago, Arigo often accomplished over a hundred medical miracles in a day. Within a few seconds and with no medical instruments, he would correctly diagnose the most complex physical ailments. Frequently he would perform painless minor surgery using no anesthetics and no antiseptic precautions. In one case, he ripped open a man's stomach with his bare hands, removed a cancerous tumor and patched the stomach together again—all this while the man was awake and smiling.[37] Dr. Henry Puharich, formerly senior research scientist at the New York University Medical Center, has filmed Arigo at work and is now studying other paranormal healers in the United States and abroad.

Throughout this chapter we have traveled along the fringes of science. We have shown how biofeedback research has encouraged underground science, and we have offered some playful speculations of our own. No one can tell for sure how far underground research will go in the next few years, but Dr. Puharich has hinted at some far-ranging possibilities. For example, at a recent lecture explaining how Arigo and other psychic healers were able to overcome the laws of medical science, he made an interesting observation: the largest concentration of U.F.O. sighting in the world centers on the space above Arigo's home.[38]

6

Biofreedom

I believe that the real impact of psychology will be felt, not through the technological products it places in the hands of powerful men, but through its effects on the public at large, through a new and different public conception of what is humanly possible and what is humanly desirable.

GEORGE A. MILLER
From his presidential address
to the American Psychological
Association

What's in a Theory?

If we were to compose a list of the most influential names of the last hundred years, certainly the name Sigmund Freud would rank near the top. His ideas have forcefully shaped present attitudes toward child rearing, sexual behavior, education, politics and even literary criticism. His influence is even more remarkable when you realize that the validity of Freudian analysis has *never* been adequately demonstrated. Some of Freud's detractors have even suggested that psychoanalysis aggravates, rather than cures, psychological problems.

The point to be made is this: a radical scientific advance has consequences far beyond the introduction of a new technique or theory. By changing prevailing conceptions of how man and his universe operate, a new scientific idea can have dramatic and unforeseen political and social consequences. Consider the case of Copernicus. His theory placing the locus of the universe not at the earth but at the sun had repercussions far beyond the province of astronomy. This new idea "did not change the productivity of the fields, turn wine into vinegar, or render less fascinating the pursuit of women or war," notes scholar Garvin McCain wryly, ". . . yet, when a long established and firmly held belief is shattered, life can never be the same." [1]

We have already speculated on the possible uses of biofeedback training. We described, among other innovations, the voluntariums of the future where people will gain control over illness through mental processes. Yet to end with these technological projections of biofeedback research would be to ignore the broader, and perhaps even more significant, implications of this new field. In an era when man increasingly relies on dogmas, drugs, authority figures and other external agents to manage his life, biofeedback training stands out as a breakthrough in self-regulation with philosophical, social and political consequences.

Biofeedback Training and Man's Self-Concept [2]

The most important impact of biofeedback training is likely to be on man's conception of himself. Ever since the Middle Ages, man's opinion of himself has been on the decline. According to early Christian thinkers, man ruled the best planet in the most important part of the universe. Then Galileo, Kepler and Copernicus came along, and the earth was demoted from central position to a mere satellite of the sun. Hardly had man recovered from this challenge to his position in space when an Englishman named Darwin began to question his status on native earth. Prior to

133

Darwin, man saw himself as the king of God's creatures, separated from the animals by the possession of a divine soul. To suggest otherwise was to invite amusement, indignation or wrath, depending upon the physical stature of the listener. Yet Darwin argued convincingly and, with the help of other scientists, prevailed. Once-proud man, relegated to the corner of the universe, was now related to his animal ancestors. It wasn't a happy family reunion. Freud delivered the final blow. Extending the notion of scientific determinism to include even human behavior, he boldly declared that all man's actions, even his noblest deeds, were the inexorable results of blind physical forces. One professor characterized the decline of man's self-concept in this way: "With Copernicus man lost his throne, with Darwin he lost his soul, and with Freud he lost his mind."

The present state of man's self-image is reflected in the daily use of such words and expressions as "alienation," "disillusionment," "loss of faith" and "lack of self-respect." The evolution of the computer, the growth in population, the ability of psychologists to control behavior with increasing accuracy—all serve to lower a person's sense of worth, his self-reliance, his sense of autonomy. The decline of man's self-concept is not an abstract or airy intellectual question. The growing problems of suicide, depression, alcoholism and drug abuse are linked to the deteriorating self-concept syndrome.

In this context, biofeedback research is an important reversal in the history of scientific psychology. For the first time, experimenters are showing that man can be the master of his own destiny rather than the slave of his juices. Every new disease, every mental state, every "involuntary" behavior brought under conscious control is a step toward revitalizing man's tarnished self-image. Not only does biofeedback training promise to enhance man's control of his internal states, it promises a renewed sense of *freedom and dignity* that will make this control both satisfying and fulfilling.

Just as biofeedback research will alter our concept of man, so it will change the institutions that serve and study him. The impact will be felt particularly in the medical sciences.

Medicine today is largely a fragmented profession. Diseases are regarded as localized phenomena, and doctors specialize accordingly. There are dermatologists for skin problems, ophthalmologists for eye disorders, gynecologists for diseases of the female reproductive organs, psychiatrists for "mental" problems and so on. Biofeedback training will be a potent force in changing this antiquated system. Each new experiment is documenting the fact that human health depends on integrated functioning of the entire organism. The "psychological" is not separate from the physical; the endocrine system does not operate independently from the autonomic and skeletal nervous systems.

In the future, doctors will be required to take a more integrated, holistic view of medicine. In addition, with the evolution of fully operational voluntariums, patients will be given greater responsibility for preventing and curing their own diseases. Indeed, medical education, instead of being concentrated in the hands of a few doctors, may become diffused within the population, perhaps even required in secondary school. Just as high-school students now take citizenship courses to improve their skills as self-governing citizens, so future schoolchildren may take biofeedback courses to improve their skills at governing the health of their own bodies.

Biofeedback training will also serve to expand the limits of medical ethics. Until now, the rigid division between "medical" and "psychological" has exempted psychologists and other social scientists from regulation by internationally accepted rules of human medical practice. However, as physical events come increasingly under mental control and vice versa, the distinction may dissolve. In particular, psychological experiments conducted on healthy subjects just to prove a theoretical point will come under closer scrutiny.

Biofeedback training will also have a dramatic impact on academic psychology. For one, the subject matter of the field will change. Brain-wave training has provided tools for the intensive investigation of consciousness, an area which, until now, has been ignored and even scorned by professionals. Kamiya anticipates the time when psychology will produce "an internal vocabulary, a language [man] can use to explain more effectively and completely how he feels inside." [3] To the extent that physical states represent levels of consciousness and awareness, "man would at last have an exact vocabulary for interpersonal communication." [4]

While biofeedback research will open new areas in psychology, it may merge others. Some scientists [5] foresee the joining of two previously unlikely allies, physiology and social psychology. Using EEG, EMG and other physiological tools, science may be able to obtain precise measures of such social phenomena as racism, friendship, religious dogmatism, conformity, political liberalism and even sexual compatibility.

The voluntary control of internal states will certainly affect other institutions, though to what extent it is almost impossible to predict. The changed character of education, for example, may extend beyond the introduction of biofeedback training courses. Students will continue to read about different cultures and historical periods, but some might choose to specialize in the study of different levels of their own consciousness. And instead of taking time off from school to have "real experiences in the outside world," students may opt for the opposite alternative: a leave of absence from normal consciousness for extended journeys into altered states of internal awareness.

Even the legal system may be touched by biofeedback research. For years lawyers have tried to resolve the conflict between the traditional legal definition of man as a free creature, responsible for his actions, and the new *zeitgeist* of social determinism, which holds that criminal actions are the unavoidable result of a faulty environment.

The implications of this conflict should not be minimized. "In recent years," notes Elmer Green, "scientists in every nation have come to realize that voluntary control of behavior is of primary importance if we hope to establish an ordered society or even maintain a society." [6] Biofeedback research may help to reintroduce the concept of volition and personal responsibility into the courtroom.

Psytocracy or Freedom?

Ever since the publication of Aldous Huxley's *Brave New World,* Americans have witnessed the transformation of gloomy science fiction into science fact. First came the operant conditioning techniques of reward and punishment, which gave any person who could read a psychology book the power to shape and control the behavior of others. "Give me a dozen healthy infants, well formed, and my own specified world to bring them up in," Dr. John Watson had boasted, "and I'll guarantee to take any one at random and train him to become any type of specialist I might select—doctor, lawyer, artist, merchant chief, and, yes, even beggar-man and thief, regardless of his talents, penchants, tendencies, abilities, vocations, and race of his ancestors." [7] Technology, marvel that it is, accelerates geometrically, so that operant conditioning was quickly followed by advances in remote control of the brain with electrical probes, genetic engineering, pharmacological alteration of personality, and the control of behavior with electronic surveillance devices.[8]

Today the situation is such that Harvard psychologist B. F. Skinner's recent book, *Beyond Freedom and Dignity,* which advocates systematic behavior control as the solution to social problems, is not only widely read but has attained best-seller status. At the 1971 meeting of the American Psychological Association, its president, Kenneth Clark, a professor at the City University of New York and a prominent liberal authority on racism and desegregation, made national headlines when he told delegates that antihostility drugs should be administered to political and

137

military leaders to "diminish their emotional propensities to respond to an international crisis by initiating a nuclear war." [9] Clark also suggested wider use of similar drugs by lesser elected officials and even civilians.

If you think this proposal is farfetched, consider that while Skinner's face was still decorating the cover of *Time* magazine a related story broke: 300,000 American schoolchildren were being given mood-changing drugs to control their behavior in the classroom.[10] In Omaha, Nebraska alone, between 5 and 10 percent of grade-school children were being controlled with medically prescribed amphetamines.

The excitement of biofeedback training is not simply its power to alter behavior. Much of what it accomplishes can also be achieved through drugs, surgery, operant conditioning and electrical stimulation of the brain. Biofeedback training is important because it places the power for change and control in the hands of the individual, not with an external authority. Of all the technologies for altering behavior, this is the first to rely on the individual's ability to guide his own destiny.

In this context, biofeedback research may even have political consequences. The United States, as well as other industrialized nations, is rapidly becoming what we have called a "psytocracy," a society that depends on externally administered methods of behavior control to maintain stability.[11] Some of these controls, such as drugs and national data banks, are obvious. Others are more subtle but no less effective. Psychologists and graduate students from Stanford University, for example, are showing teachers in local high schools how to run classes with conditioning methods. High-school students are unknowingly manipulated into "obedient" and "proper" behavior patterns with systematic presentations of rewards and punishments. Commenting on the prospect of a psychologically controlled society, Professor Perry London of the University of Southern California wrote recently, "As 1984 draws near, it appears that George Orwell's . . . conceits of the technology by which tyranny could impress its will upon men's minds were much too modest. By that time, the means at hand will be more sophisticated and efficient than Orwell ever dreamed, and they will be in at least modest use, as they have already begun to be." [12]

Technology per se is not to blame. We never have to use it if we don't want to. The problem is the conviction that the control of each person's behavior by an external authority is the only effective way to manage society. This is Skinner's belief. It is Clark's belief. It is also an assumption held both consciously and unconsciously by a growing number of scientists, administrators and politicians.

"The serious threat to our democracy," John Dewey once wrote, ". . . is the existence within our own personal attitudes and within our own institutions of conditions similar to those which have given a victory to external authority, discipline, uniformity and dependence. . . . The battlefield is also accordingly here—within ourselves and our institutions." [13]

If psytocracy is to be avoided, people must come to believe—not merely hope or wish—that man is more than a biological automaton, that he is capable of controlling himself, by himself and for himself. People must believe that man has the intuitive wisdom to move in ways that are beneficial both for himself and for others. Biofeedback research is the first scientific enterprise to validate these optimistic beliefs with empirical evidence. Even if biofeedback training does not develop as rapidly as we anticipate, its value to our democracy—indeed, to the perpetuation of the free spirit everywhere—will still be incalculable.

NOTES

INTRODUCTION

1. "Trained control of bodily states might well be added to the [educational] curriculum, perhaps beginning as early as elementary school," Dr. Joe Kamiya predicts. His forecast appears in a 1968 article that first focused public attention on biofeedback research. (Kamiya, J. Conscious control of brain waves. *Psychology Today*, 1968, 1, p. 60.)

2. Kiefer, D. Meditation and biofeedback. In: J. White (ed.). *The Highest State of Consciousness*. New York: Doubleday (Anchor Original), 1972, p. 329.

3. One investigator, for example, became so proficient at controlling his brain waves that he used them to send Morse code messages (with a little help from an electroencephalograph machine and a computer). See: Dewan, E. Communication by voluntary control of the EEG. *Proceedings of the Symposium of Biomedical Engineering*, 1966, 1, 349-351. Another person, composer David Rosenboom, regulates his mental states to create brain wave music! Mr. Rosenboom, the composer of the seventy-two-hour piece "How Much Better If Plymouth Rock Had Landed on the Pilgrims," reported to the *New York Times* (November 25, 1970) that he composed by controlling the emission of certain brain waves and making them into music. He recently gave a brain-wave concert at New York's Automation House.

4. Smart, A. Conscious control of physical and mental states. *Menninger Perspective*, April-May, 1970.

5. To fully appreciate this argument, you must know something about the character of academic psychology in the United States. At the turn of the century, American psychologists were looking enviously at their colleagues in the physics and chemistry departments. These "hard scientists" had discovered laws that could be observed, tested and written up in the journals. In contrast, psychology was little more than a weird mixture of philosophy, mysticism and religion. In an effort to make psychology a "legitimate" science, the behaviorists excised all references to freedom, self-control, volition, consciousness and so on from academic psychology. "From now on," they declared, "psychology will deal only with the

140

deterministic laws of human behavior." Today psychology is divided between the dominant behaviorists, on the one hand, and a resurgent humanist movement, on the other.

6. Ornstein, R. Personal communication, 1971.

7. Mulholland, T. Can you really turn on with alpha? Paper presented to the Massachusetts Psychological Association meeting, May 7, 1971, p. 1 (mimeo report).

CHAPTER 1: Understanding Biofeedback

1. Mayr, O. The origins of feedback control. *Scientific American,* 1970, 223, p. 111.

2. *Ibid.*

3. Hardyck, C., and Petrinovich, L. Treatment of subvocal speech during reading. *Journal of Reading,* 1969, 12, 361-368+.

4. *Ibid.,* p. 367.

5. Paraphrase of the following article title: Lang, P. Autonomic control or learning to play the internal organs. *Psychology Today,* October, 1970, 37-41+.

6. Rorvik, D. The wave of the future: Brain waves. *Look,* October 6, 1970, p. 91.

7. The authors first heard this term from Adam Crane of Toomim Bio-Feedback Laboratories in New York City.

8. Criswell, E. Personal communication, 1971.

9. Miller, N., et al. Learned modifications of autonomic functions: A review and some new data. *Circulation Research* (Supplement 1), 1970, 26-27, p. I-3.

10. DiCara, L., & Miller, N. Instrumental learning of vasomotor responses by rats: Learning to respond differentially in the two ears. *Science,* 1968, 159, 1485-1486.

11. For a further discussion of this topic, please see Chapter 6.

CHAPTER 2: Biofeedback Training for Your Health

1. Phillips, D. Dying as a form of social behavior. Doctoral dissertation, Princeton University, 1969, p. 164.

2. It seems there is also a "will to die." See, for example, Eisendrath, R. The role of grief and fear in the death of kidney transplant patients. *American Journal of Psychiatry,* 1969, 126, 381-387.

3. Lowinger, P., and Dobie, S. What makes the placebo work? *Archives of General Psychiatry*, 1969, 20, p. 84 (a good description of how long the placebo effect has been with us).

4. Murphy, G. Psychology in the year 2000. *American Psychologist*, 1969, 24, pp. 525-526.

5. In Chapter 4 (EMG machine section) we will elaborate on this point further.

6. Grim, P. Anxiety change produced by self-induced muscle tension and by relaxation with respiration feedback. *Behavior Therapy*, 1971, 2, p. 11.

7. A control group that was also told to relax but received amplifier background hum rather than respiration feedback did not show a similar reduction in anxiety.

8. Luce, G., and Peper, E. Biofeedback: Mind over body, mind over mind. *New York Times Magazine*, September 12, 1971, p. 132.

9. *Ibid*.

10. Budzynski, T., and Stoyva, J. Biofeedback techniques in behavior therapy and autogenic training. Unpublished manuscript, University of Colorado Medical Center, 1971.

11. Collier, B. Brain power: The case for bio-feedback training. *Saturday Review*, April 10, 1971, p. 58.

12. Petroni, L. Personal communication, 1971.

13. *Ibid*.

14. *Ibid*.

15. Recently, her efforts have received TV and press coverage in the Tucson area.

16. Sargent, J., Green, E., and Walters, E. Preliminary report on the use of autogenic feedback techniques in the treatment of migraine and tension headaches. Unpublished manuscript, Menninger Foundation, 1971, p. 2.

17. *Ibid*., p. 7.

18. Budzynski, T., Stoyva, J., and Adler, C. Feedback-induced muscle relaxation: Application to tension headaches. *Journal of Behavior Therapy and Experimental Psychiatry*, 1970, 1, 205-211.

19. Luce and Peper, Biofeedback . . . , p. 132.

20. Of all the medical problems currently being treated with biofeedback training, work in headache control is most advanced.

21. Shapiro, D., et al. Effects of feedback and reinforce-

ment on the control of human systolic blood pressure. *Science,* 1969, 163, 588-590.

22. Brener, J., and Kleinman, R. Learned control of decreases in systolic blood pressure. *Nature,* 1970, 226, p. 1063.

23. Christy, A., and Vitale, J. Operant conditioning of high blood pressure: A pilot study. Unpublished manuscript, San Francisco Veterans Administration Hospital, 1971. The authors define labile hypertensives as "those patients who show wide variations of both systolic and diastolic blood pressure, particularly in response to emotional stress. Essential hypertensives are patients whose systolic and diastolic pressures remain elevated unless medicated; the etiology is unknown." In this particular study, results were inconclusive for essential hypertensives.

24. Luce and Peper, p. 136.

25. Pearlman, C., and Greenberg, R. Medical-psychological implications of recent sleep research. *Psychiatry in Medicine,* 1970, 1, 261-276.

26. They used the procedure we discussed in the treatment of anxiety.

27. Luce and Peper, p. 132.

28. Sterman, M., Howe, R., and MacDonald, L. Facilitation of spindle-burst sleep by conditioning of electroencephalographic activity while awake. *Science,* 1970, 167, 1146-1148.

29. Luce and Peper, p. 132.

30. *Ibid.*

31. Booker, H., Rubow, R., and Coleman, P. Simplified feedback in neuromuscular retraining: An automated approach using electromyographic signals. *Archives of Physical Medicine and Rehabilitation,* 1969, 50, p. 625.

32. This description originates with Dr. Peter J. Lang.

33. Some investigators would still take issue with this point of view, claiming that man might be exerting such regulation by controlling muscle groups that have always been under his voluntary control. We (and most of our colleagues) doubt this—but until a person is willing to try cardiac control under curare we won't know for sure!

34. Weiss, T., and Engel, B. Operant conditioning of heart rate in patients with premature ventricular contractions. *Psychosomatic Medicine,* 1971, 33, p. 319.

35. Collier, p. 11.

36. *Newsweek,* June 21, 1971, p. 62.

37. Weiss and Engel, p. 320.

38. Luce and Peper, p. 134. For information on the second study with heart patients see Engel, B., and Melmon, K. Operant conditioning of heart rate in patients with cardiac arrhythmias. *Conditional Reflex*, 1968, 3, 130 (abstract).

39. Green, E. Personal communication, 1971.

40. *Ibid.*

41. Whatmore, G., and Kohli, D. Dysponesis: A neurophysiologic factor in functional disorders. *Behavioral Science*, 1968, 13, 102-124.

42. *Ibid.*, p. 115.

43. *Ibid.*, p. 121.

44. Budzynski, T., and Stoyva, J. An instrument for producing deep muscle relaxation by means of analog information feedback. *Journal of Applied Behavior Analysis*, 1969, 2, 231-237.

45. This section draws heavily from Gardner Murphy's paper presented at the First Annual Meeting of the Bio-Feedback Research Society, October, 1969. It is reprinted in: T. Barber, et al. (eds.). *Biofeedback and Self-Control, 1970.* Chicago: Aldine-Atherton, 1971.

46. *Ibid.*, p. 491.

47. *Ibid.*, p. 493.

48. Mulholland, T. Feedback electroencephalography. *Activitas Nervosa Superior* (Prague), 1968, 10, p. 436.

49. Budzynski and Stoyva, Biofeedback techniques . . . , pp. 1-38.

50. Mulholland, T., and Benson, D. Feedback electroencephalography in clinical research. Unpublished manuscript, Bedford, Massachusetts, Veterans Administration Hospital, 1971.

51. Kamiya, J. Conscious control of brain waves. *Psychology Today*, 1968, 1, p. 60.

52. Toomim, M. Personal communication, 1971.

53. Peper, E. Personal communication, 1971.

54. Korein, J. Personal communication, 1971.

55. Jacobs, A., and Felton, G. Visual feedback of myoelectric output to facilitate muscle relaxation in normal persons and patients with neck injuries. *Archives of Physical Medicine and Rehabilitation*, 1969, 50, 34-39.

56. Malmo, R. Emotions and muscle tension: The story of Anne. *Psychology Today*, March, 1970, 64-67+.

144

57. Muscle is retrained at home. *Medical World News,* December 10, 1971, 35.

58. Peper, E. Personal communication, 1971.

59. Shaffer, R. Bio-Feedback. *Wall Street Journal,* April 19, 1971.

60. Green E., Green, A., and Walters, E. Voluntary control of internal states: Psychological and physiological. *Journal of Transpersonal Psychology,* 1970, 2, p. 23.

61. Green, E. Personal communication, 1971.

CHAPTER 3: The Inner Trip

1. James, W. *The Varieties of Religious Experience.* New York: Mentor Books, 1958, p. 292.

2. The borders between brain states are somewhat arbitrary. Some researchers, for example, will define alpha as seven to fourteen cycles per second.

3. This is current speculation, not established fact.

4. Kamiya, Conscious control . . . , p. 57.

5. *Ibid.*

6. *Ibid.,* p. 58.

7. Kamiya, J. Operant control of the EEG alpha rhythm and some of its reported effects on consciousness. In: C. Tart (ed.). *Altered States of Consciousness.* New York: Wiley, 1969, p. 515.

8. Criswell, E. Personal communication, 1971.

9. Luce and Peper, p. 138.

10. Green, E. Personal communication, 1971.

11. Mulholland, Can you really . . . , p. 6.

12. Toomim, M. Personal communication, 1971.

13. Luce and Peper, p. 136.

14. Walter, W. The social organ. *Impact of Science on Society,* 1968, 18, p. 184.

15. Gould, D. All a matter of brain waves. *New Statesman,* January 17, 1969, p. 77.

16. Hoebel, B. Personal communication, 1968.

17. *Newsweek,* June 21, 1971, p .65.

18. Bakan, P. The eyes have it. *Psychology Today,* April, 1971, p. 64.

19. *Ibid.,* p. 67.

20. *Ibid.,* p. 96.

21. Ornstein, R. Personal communication, 1971.

22. *Ibid.*

23. Hoenig, J. Medical research on yoga. *Confinia Psychiatrica,* 1968, 11, p. 74.

24. Luce and Peper, p. 132.

25. For a discussion of meditation and the reduction of mental and physical tension see Wallace, R. Physiological effects of transcendental meditation. *Science,* 1970, 167, p. 1754. Meditation and pain endurance is examined in Anand, B., et al. Some aspects of electroencephalographic studies in Yogis (*Electroencephalography and Clinical Neurophysiology,* 1961, 13, p. 453). The relationship between meditation and the development of empathy is reviewed by T. Lesh in his article Zen meditation and the development of empathy in counselors. In: T. Barber et al. (eds.). *Biofeedback and Self-Control, 1970,* 113-148. For a recent article on physiological changes that take place during meditation, we refer you to Wallace, R., et al. A wakeful hypometabolic physiologic state. *American Journal of Physiology,* 1971, 221, 795-799.

26. Quoted in *Time,* October 25, 1971, p. 51. These findings were also reported in the hearings before the House of Representatives Select Committee on Crime (June 2-4, 1971). In the report, Dr. Benson wrote to the committee chairman, "A study of 1,862 subjects indicated that individuals who regularly practiced transcendental meditation (a) decreased or stopped abusing drugs, (b) decreased or stopped engaging in drug selling activity, and (c) changed their attitudes in the direction of discouraging others from abusing drugs."

27. See, for example, Wallace, R. Physiological effects of transcendental meditation. *Science,* 1970, 167, p. 1754. Because meditation is associated with slower heart rate, lower respiration and other physiological correlates that can be measured objectively, it is becoming an attractive form of psychotherapy. In the words of University of California psychologist Joe Hart, "Meditation appears to be the only training technique known to psychology for which there are clearly demonstrable physiological indicators of the training's success or failure."

28. Kasamatsu, A., and Hirai, T. An electroencephalographic study of the Zen meditation (Zazen). In: C. Tart (ed.). *Altered States of Consciousness,* pp. 489-501.

29. Anand, B., Chhina, G., and Singh, B. Some aspects of

electroencephalographic studies in Yogis. *Electroencephalography and Clinical Neurophysiology*, 1961, 13, 452-456.

30. Stoyva, J. The public (scientific) study of private events. In: T. Barber et al. (eds.). *Biofeedback and Self-Control, 1970*, p. 35.

31. Rorvik, D., The wave. . . , p. 91.

32. Green, Green and Walters, Voluntary control. . . , p. 8.

33. Green, E. Personal communication, 1971.

34. Green, Green and Walters, Voluntary control. . . , pp. 12-20.

35. Green, E., Green, A., and Walters E. Self-regulation of internal states. In: J. Rose (ed.). *Progress of Cybernetics: Proceedings of the International Congress of Cybernetics, London, 1969.* London: Gordon & Breach, 1970, p. 1316.

36. Personal communication, 1971.

37. August, R. Hallucinatory experiences utilized for obstetric hypnoanesthesia. *American Journal of Clinical Hypnosis*, 1960, 3, 90-94.

38. Some scientists, led by Theodore X. Barber, claim that hypnosis cannot technically be regarded as a state of consciousness. Dr. Ernest Hilgard of Stanford University and others disagree.

39. August, p. 90.

40. Barber, T. Physiological effects of hypnosis and suggestion. In: Barber, *Biofeedback*. . . , pp. 188-256.

41. Hilgard, E. Personal communication, 1971.

42. Hart, J. In: Barber, *Biofeedback*. . . , p. 57.

43. Galbraith, G., et al. EEG and hypnotic susceptibility. *Journal of Comparative and Physiological Psychology*, 1970, 1, 125-131.

44. Engstrom, D., London, P., and Hart, J. Hypnotic susceptibility increased by EEG alpha training. *Nature*, 1970, 227, 1261-1262.

45. *Ibid.* A recent study also suggests that EMG biofeedback might be used to enhance hypnotic susceptibility (see bibliography for Wickramasekera, I. Effects of EMG. . .).

CHAPTER 4: What to Do till the Revolution Comes

1. Luce and Peper, p. 139.

2. Actually, there are other machines as well (e.g., GSR—galvanic skin response—instruments), but these either have

more limited applications, are not as well perfected at this time or would not interest many readers. In the coming years a great many more different machines will be on the market, judging from current production plans of major companies we talked with.

3. Green, E. Personal communication, 1971.

4. Steiner, S. Personal communication, 1971.

5. Cambridge Cyborgs and New York Cyborgs are the same company located in different cities.

6. This fine machine is also more costly ($650) than those produced by the Cyborg or Toomim companies. Also, the president of Scott Electronics recommends that if you wish to purchase one of their machines you do so through a doctor or scientist in your area.

7. Steiner, S. Personal communication, 1971.

8. Criswell, E. Personal communication, 1971.

9. Shapiro, D., and Schwartz, G. Psychophysiological contributions to social psychology. *Annual Review of Psychology,* 1970, 21, p. 91.

10. Toomim, H. Personal communication, 1971.

11. We also strongly recommend against using any kind of EEG machine that plugs directly into a wall socket (or includes any instruments that attach to it and plug into a wall socket). As of this writing we know of no such machine on the consumer market (all the ones we are familiar with use battery power). The reason for this recommendation is the unlikely but possible chance that you might receive an electric shock from an improperly working, plug-in-model EEG apparatus.

12. Budzynski and Stoyva. Biofeedback techniques. . . , pp. 20-21.

13. Miller, N. Learning of visceral and glandular responses. *Science,* 1969, 163, p. 444.

14. Scott, K. Personal communication, 1971.

15. If you decide to ask this question you might want to check the brand name of the machinery being utilized against our recommendations. This question will become most useful if and when a regulatory body decides to evaluate and control biofeedback equipment (you can see how good your organization's machine stacks up against others, as determined by impartial government survey).

16. Everett, A. Personal communication, 1971.

17. Green, E. Personal communication, 1971.

18. The authors gratefully record their indebtedness to Gay Luce, author of the book *Body Time*, for her suggestions concerning this section of the text.

19. Luce, G. Understanding body time in the 24-hour city. *New York Magazine*, November 15, 1971, p. 40.

20. *Ibid.*, p. 42.

21. Bernal, M. Behavioral feedback in the modification of brat behaviors. *Journal of Nervous and Mental Disease*, 1969, 148, 375-385.

22. *Ibid.*, p. 384.

23. *Ibid.*, p. 379.

24. Stoller, F. The long weekend. *Psychology Today*, December, 1967, p. 32.

25. Holzman, P. *Psychology Today*, November, 1971, p. 98.

26. *Ibid.*

27. McFall, R. Effects of self-monitoring on normal smoking behavior. *Journal of Consulting and Clinical Psychology*, 1970, 35, p. 135.

28. Wendell, C. Personal communication, 1971.

29. Stoller, The long weekend, p. 33.

30. We refer, of course, to B. F. Skinner. He presents his argument most brazenly in his book *Walden Two* (New York: Macmillan 1948). Although it is labeled a work of fiction, *Walden Two* is, in reality, a reflection of Skinner's scientific thinking from start to finish—an application of his operant conditioning methods to design a society created and governed by psychologists.

CHAPTER 5: Underground Science

1. We acknowledge our debt to Jim Beal of NASA for alerting us to this quote.

2. Gordon, T. Bucking the scientific establishment. *Playboy*, April, 1968, p. 134.

3. Tiller, W. Personal communication, 1971.

4. Vogel, M. Personal communication, 1971.

5. Putoff, H. Personal communication, 1971.

6. Tart, C. Personal communication, 1971.

7. See: Aserinsky, E., and Kleitman, N. Regularly occurring periods of eye motility and concomitant phenomena

during sleep. *Science,* 1953, 118, 273-274. Also, Dement, W., and Kleitman, N. Cyclic variations in EEG during sleep and their relations to eye movements, body mobility and dreaming. *Electroencephalography and Clinical Neurophysiology,* 1957, 9, 673-690.

8. Material relating to this distinction is drawn from the authors' *Requiem for Democracy?* New York: Holt, Rinehart & Winston, 1971.

9. See, for example, Dewan, E. Communication by voluntary control of the EEG. *Proceedings of the Symposium of Biomedical Engineering,* 1966, 1, 349-351.

10. Schafer, W. Further developments of the field effect monitor. Life Sciences, Corvair Division of General Dynamics, Report GDC-ERR-AN-1114, October, 1967.

11. *Science Digest,* June, 1968, p. 88.

12. Kamiya, Conscious control. . . , p. 59.

13. Beal, J. Personal communication, 1971.

14. Tiller, W. Personal communication, 1971.

15. Targ, R. Personal communication, 1971.

16. Targ, R. Learning clairvoyance and precognition with an extrasensory perception teaching machine. Paper presented at the Parapsychology Association Meeting, Durham, North Carolina, 1971, p. 7 (mimeo report).

17. Schmeidler, G. Respice, adspice, prospice. Presidential address, Parapsychological Association, 1971, p. 5 (mimeo report).

18. Letter to Dr. J. B. Rhine, quoted in: Beal, J. Paraphysics and parapsychology: Recent developments associated with bioelectric field effects. Unpublished manuscript, 1971, p. 11.

19. Tiller, W. The A.R.E. visit to Russia. Unpublished manuscript, undated.

20. Harrell, M. Personal communication, 1971. It has also been recently reported (*Time,* January 10, 1972, p. 48) that certain kinds of "accident rates . . . are influenced by phases of the moon, solar cycles and other natural phenomena."

21. Luce, G. *Body Time.* New York: Pantheon, 1971.

22. Tiller, W. Personal communication, 1971.

23. Beal, Paraphysics. . . , p. 4.

24. Beal, Personal communication, 1971.

25. Beal, Paraphysics. . . , pp. 4-5.

26. *Ibid.,* p. 5.

27. This idea was suggested to us by Dr. Barbara Brown.

28. Beal, p. 5. For further discussion of electrosleep, see the bibliography listings for Lippold et al., Ramsay, et al., and Rosenthal et al.

29. Backster, C. Evidence of a primary perception in plant life. *International Journal of Parapsychology*, Winter, 1968, 329-348.

30. McGraw, W. Plants are only human. *Argosy*, June, 1969, p. 25. For those interested in trying their hand at plant communication, a description of the machinery is published in: L. Lawrence. Electronics and the living plant. *Electronics World*, October, 1969, 25-28.

31. Vogel, M. Personal communication, 1971.

32. See, for example, Loehr, F. *The Power of Prayer on Plants*. New York: New American Library (Signet Books), 1969.

33. Vogel, M. Personal communication, 1971.

34. Pearce, J. *The Crack in the Cosmic Egg*. New York: Julian Press, 1971.

35. *Ibid.*, p. 101.

36. *Ibid.*, p. 106. Pearce also comments: "A high point was reached when one of the fakirs noticed a professor of psychology avidly intrigued and dumbfounded. The fakir, sensing the longing, told the good professor he, too, could walk the fire if he so desired—*by holding the fakir's hand*. The good man was seized with faith that he could, shed his shoes, and hand-in-hand they walked the fire ecstatic and unharmed."

37. From a lecture by Dr. Puharich at De Anza College, California, October 30, 1971.

38. *Ibid.*

CHAPTER 6: Biofreedom

1. McCain, G., and Segal, E. *The Game of Science*. Belmont, Calif.: Brooks-Cole, 1969, p. 5.

2. Material relating to this section is drawn (in condensed form) from the authors' *Requiem for Democracy?*

3. *Time*, July 18, 1969, p. 67.

4. Kamiya, Conscious control. . . , p. 60.

5. Shapiro and Schwartz, Psychophysiological contributions. . . , pp. 87-88.

6. Green, Green and Walters, Voluntary control. . . , p. 2.

7. Watson, J. *Behaviorism*. New York: People's Institute, 1924, p. 82.

8. A broad survey of behavior control techniques is provided in *Requiem for Democracy?* (Ch. 1).

9. Quoted from Kenneth Clark's presidential address to the American Psychological Association convention (September, 1971).

10. Witter, C. Drugging and schooling. *Transaction*, July-August, 1971, 30-34. Witter states on p. 30: "A careful reading of Department of Health, Education, and Welfare (HEW) testimony at the Gallagher hearing suggests that 200,000 children in the United States are now being given amphetamine and stimulant therapy, with probably another 100,000 receiving tranquilizers and anti-depressants." For a related article see Rogers, J. Drug abuse—just what the doctor ordered. *Psychology Today*, September, 1971, p. 20 (the Omaha, Nebraska, figure will be found here).

11. The term "psytocracy" is a neologism coined by the authors, first appearing in *Requiem for Democracy?*

12. London, P. *Behavior Control*. New York: Harper & Row, 1969, pp. 7-8.

13. Dewey, J. *Freedom and Culture*. New York: Putnam, 1939, p. 49.

ANNOTATED LIST OF SUGGESTED READINGS

CHAPTER 1. Understanding Biofeedback

Barber, T., et al. (eds.). *Biofeedback and Self-Control, 1970.* Chicago: Aldine-Atherton, 1971.

A collection of articles on biofeedback by various investigators in the field. This is an annual; thus in March of 1972 the 1971 volume of *Biofeedback and Self-Control* is due for publication. The editors did a nice job of sampling research from the field in the 1970 edition.

Black, A. The operant conditioning of central nervous system electrical activity. In: G. Bower (ed.). *The Psychology of Learning and Motivation* (Vol. 6), New York: Academic Press 1972, in press.

A thoughtful and provocative presentation by a researcher in the field.

Collier, B. Brain power: The case for bio-feedback training. *Saturday Review,* April 10, 1971, 10+.

Interesting article highlighted by interviews with Dr. Fehmi, Dr. Miller and David Rosenboom, the man who recently gave a "brain-wave concert" at New York's Automation House.

DiCara, L. Learning in the autonomic nervous system. *Scientific American,* 1970, 222, 30-39.

A lucid presentation by one of the pioneering investigators in the field. Emphasis is on animal research.

Kamiya, J. Conscious control of brain waves. *Psychology Today,* 1968, 1, 56-60.

This is the article that first focused public attention on biofeedback research. It is highly readable but limited to a discussion of brain wave biofeedback that is somewhat dated.

Katkin, E., and Murray, E. Instrumental conditioning of autonomically mediated behavior: Theoretical and methodological issues. *Psychological Bulletin,* 1968, 70, 52-68.

Read in conjunction with: Crider, A., Schwartz, G., and Shnidman, S. On the criteria for instrumental autonomic conditioning: A reply to Katkin and Murray, *Psychological Bulletin*, 1969, 71, 455-461. And: Katkin, E., Murray, E., and Lachman, R. Concerning instrumental autonomic conditioning: A rejoinder. *Psychological Bulletin*, 1969, 71, 462-466.

Academic researchers locked in theoretical battle. A few bloodied noses—but biofeedback research goes on.

Lang, P. Autonomic control or learning to play the internal organs. *Psychology Today*, October, 1970, 37+.

Well written but somewhat limited article as far as material covered is concerned. Lang's own work in cardiac control is highlighted.

Luce, G., and Peper, E. Biofeedback: Mind over body, mind over mind. *New York Times Magazine*, September 12, 1971, 34+.

A superb, thoughtful piece that provides a complete overview of biofeedback research. By far the best popular article on biofeedback written to date.

Miller, N. Learning of visceral and glandular responses. *Science*, 1969, 163, 434-445.

The man responsible for the "greeting of the nervous system" describes his revolutionary program of research. A fine presentation.

Rorvik, D. The wave of the future: Brain waves. *Look*, October 6, 1970, 88+.

Clearly written, but depends too much on quotes from Dr. Kamiya and Dr. Brown.

Shaffer, R. Bio-Feedback. *Wall Street Journal*, April 19, 1971.

A fine piece of journalism; for a brief article it is one of the most complete overviews we have.

Smart, A. Conscious control of physical and mental states. *Menninger Perspective*, April-May, 1970.

Fine description of Dr. Elmer and Alyce Green's work at the Menninger Foundation.

Turning on with alpha waves. *Life*, August 21, 1970, 60-61.

Very short article with some nice pictures.

Two brief but informative articles on biofeedback have also appeared in *Time* (Controlling the inner man, July 18, 1969, and Alpha wave of the future, July 19, 1971).

CHAPTER 2. Biofeedback Training for Your Health

Basmajian, J. Control and training of individual motor units. *Science*, 1963, 141, 440-441.

A pioneering piece of research using biofeedback for the control of individual motor units. For other examples of this type of research see: Gray, E. Conscious control of motor units in a tonic muscle. *American Journal of Physical Medicine*, 1971, 50, 34-40. Also: Marsden, C., Meadows, J., and Hodgson, H. Observations on the reflex response to muscle vibration in man and its voluntary control. *Brain*, 1969, 92, 829-846.

Benson, H., Shapiro, D., Tursky, B., and Schwartz, G. Decreased systolic blood pressure through operant conditioning techniques in patients with essential hypertension. *Science*, 1971, 173, 740-742.

One of the three outstanding published studies on biofeedback training and hypertension.

Block, J., Lagerson, J., Zohman, L., and Kelly, G. A feedback device for teaching diaphragmatic breathing. *American Review of Respiratory Disease*, 1969, 100, 577-578.

Brief but important article in an area of investigation receiving increasing attention by biofeedback scientists.

Booker, H., Rubow, R., and Coleman, P. Simplified feedback in neuromuscular retraining: An automated approach using electromyographic signals. *Archives of Physical Medicine and Rehabilitation*, 1967, 50, 621-625.

Biofeedback training used to help a middle-aged woman rehabilitate facial muscles damaged in an automobile accident.

Brener, J., and Kleinman, R. Learned control of decreases in systolic blood pressure. *Nature*, 1970, 226, 1063-1064.

One of the three outstanding published studies on biofeedback training and hypertension.

Budzynski, T., Stoyva, J., and Adler, C. Feedback-induced muscle relaxation: Application to tension headaches. *Journal of Behavior Therapy and Experimental Psychiatry*, 1970, 1, 205-211.

Milestone article by one of the leading investigators in the field. For the treatment of migraine headaches, see Sargent et al.

Engel, B., and Melmon, K. Operant conditioning of heart rate in patients with cardiac arrhythmias. *Conditional Reflex,* 1968, 3, 130.

One of two published studies using heart-rate biofeedback training to help patients with heart trouble. For the other study, See Weiss and Engel.

Grim, P. Anxiety change produced by self-induced muscle tension and by relaxation with respiration feedback. *Behavior Therapy,* 1971, 2, 11-17.

Anxiety reduction through respiration biofeedback.

Hardyck, C., and Petrinovich, L. Treatment of subvocal speech during reading. *Journal of Reading,* 1969, 12, 361-368+.

EMG biofeedback training utilized to help individuals overcome "subvocalization," the tendency to silently mouth words while reading. Such a habit limits reading speed to a ceiling of about 150 words per minute while increasing reader fatigue.

Jacobs, A., and Felton, G. Visual feedback of myoelectric output to facilitate muscle relaxation in normal persons and patients with neck injuries. *Archives of Physical Medicine and Rehabilitation,* 1969, 50, 34-39.

Another good example of the value of EMG feedback in the treatment of medical problems.

Lang, P., and Melamed, B. Case report: Avoidance conditioning therapy of an infant with chronic ruminative vomiting. *Journal of Abnormal Psychology,* 1969, 74, 1-8.

Dramatic story of how EMG feedback, in conjunction with behavior modification, is used to save a young child's life.

Malmo, R. Emotions and muscle tension: The story of Anne. *Psychology Today,* March, 1970, 64-67+.

EMG feedback used in the treatment of hysterical deafness.

Murphy, G. Experiments in overcoming self-deception. In: T. Barber et al. (eds.). *Biofeedback and Self-Control.*

Using biofeedback to overcome self-deception.

Muscle is retrained at home. *Medical World News,* December 10, 1971, 35.

Exciting report of EMG biofeedback utilized to help stroke victims regain use of paralyzed muscles.

Sargent, J., Green, E., and Walters, E. Preliminary report on the use of autogenic feedback techniques in the treatment of migraine and tension headaches. Unpublished manuscript, Menninger Foundation, 1971.

    This crucial article should soon be published and more readily available to the interested reader. The treatment described here is for the treatment of migraine headaches. For the treatment of tension headaches, see Budzynski et al.

Shapiro, D., Tursky, B., Gershon, E., and Stern, M. Effects of feedback and reinforcement on the control of human systolic blood pressure. *Science,* 1969, 163, 588-590.

    One of the three outstanding published studies on biofeedback training and hypertension.

Weiss, T., and Engel, B. Operant conditioning of heart rate in patients with premature ventricular contractions. *Psychosomatic Medicine,* 1971, 33, 301-321.

    One of two published studies using heart-rate biofeedback training to help patients with heart trouble. This is the major investigation. For the other study, see Engel and Melmon.

Whatmore, G., and Kohli, D. Dysponesis: A neurophysiologic factor in functional disorders. *Behavioral Science,* 1968, 13, 102-124.

    A treatment based on EMG-type procedures.

CHAPTER 3. The Inner Trip

Brown B. Recognition of aspects of consciousness through association with EEG alpha activity represented by a light signal. *Psychophysiology,* 1970, 6, 442-452.

    Studying mental states through brain-wave biofeedback.

Henahan, D. Music draws strains direct from brains. *New York Times,* November 25, 1970.

    Delightful article on how composer David Rosenboom uses brain-wave control to compose "brain-wave music."

Kamiya, J. Operant control of the EEG alpha rhythm and some of its reported effects on consciousness. In: C. Tart (ed.). *Altered States of Consciousness.*

    A look at alpha brain-wave control by the man who

started the whole ball rolling. Another good article by the same author, already mentioned, is: Conscious control of brain waves. *Psychology Today*, 1968, 1, 56-60.

Lynch, J., and Paskewitz, D. On the mechanisms of the feedback control of human brain wave activity. *Journal of Nervous and Mental Disease*, 1971, 153, 205-217.

A critical look at the factors underlying brain-wave control. Raises some interesting questions, but will be hard reading for the nonscientist.

Nowlis, D., and Kamiya, J. The control of electroencephalographic alpha rhythms through auditory feedback and the associated mental activity. *Psychophysiology*, 1970, 6, 476-484.

Another paper describing brain-wave control in the laboratory and the associated mental states.

Mulholland, T. Can you really turn on with alpha? Paper presented at the meeting of the Massachusetts Psychological Association, May 7, 1971.

Taking dead aim on the "cult of the alpha high," Dr. Mulholland doesn't pull any punches in his condemnation of those who see alpha control as a nirvana state. A sobering presentation for those who don't appreciate the limits of alpha brain-wave control.

Rational vs. Aesthetic Consciousness

Bakan, P. The eyes have it. *Psychology Today*, April, 1971, 64+.

Well-written easy-to-read article on the author's research concerning functional differences between the cerebral hemispheres.

Dimond, S., and Beaumont, G. Use of two cerebral hemispheres to increase brain capacity. *Nature*, 1971, 232, 270-271.

Exciting evidence for the "two brains" theory.

Biofeedback and Meditation

Kasamatsu, A., and Hirai, T. An electroencephalographic study on the Zen meditation (Zazen). In: C. Tart (ed.). *Altered States of Consciousness*, pp. 489-501.

Studying meditators' brain-wave patterns.

Kiefer, D. Meditation and bio-feedback. In: J. White (ed.). *The Highest State of Consciousness*.

One man's description of the relationship between meditation and biofeedback.

Mysterious Theta

Green, E., Green, A., and Walters, E. Voluntary control of internal states: Psychological and physiological. *Journal of Transpersonal Psychology* 1972, 2, 1-26.

The best presentation available of evidence linking theta and creativity. See pp. 12-22.

Hypnotic Breakthrough

Galbraith, G., London, P., Leibovitz, M., Cooper, L., and Hart, J. EEG and hypnotic susceptibility. *Journal of Comparative and Physiological Psychology,* 1970, 72, 125-131.

Finding a relationship between hypnotic susceptibility and specific brain-wave states.

Engstrom, D., London, P., and Hart, J. Hypnotic susceptibility increased by EEG alpha training. *Nature,* 1970, 227, 1261-1262.

Important evidence suggesting a link between hypnotic susceptibility and operant alpha rhythms.

Wickramasekera, I. Effects of EMG feedback training on susceptibility to hypnosis: Preliminary observations. *Proceedings,* 79th Annual American Psychological Association Convention, 1971.

The use of EMG biofeedback training to increase susceptibility to hypnosis—the first published finding of its kind.

CHAPTER 4. What to Do till the Revolution Comes

Body Charting

Luce, G. *Body Time.* New York: Pantheon, 1971.

The most complete and important work on the topic of body rhythms to date.

————. Understanding body time in the 24-hour city. *New York Magazine,* November 15, 1971, 37+.

The importance of body time for the city dweller.

Different Kinds of Feedback

Bernal, M. Behavioral feedback in the modification of brat behaviors. *Journal of Nervous and Mental Disease,* 1969, 148, 375-385.

The use of visual feedback to change behavior.

Stoller, F. The long weekend. *Psychology Today*, December, 1967. 28-33.

Another example of visual feedback and its power to help people change their ways of behaving.

McFall, R. Effects of self-monitoring on normal smoking behavior. *Journal of Consulting and Clinical Psychology*, 1970, 35, 135-142.

An example of paper-and-pencil feedback and its impact on human actions.

CHAPTER 5. Underground Science

What You Won't Read in the Journals

Gordon, T. Bucking the scientific establishment. *Playboy*, April, 1968, 127+.

What happens to unorthodox theorists like Immanuel Velikovsky, James McConnell and Albert Schatz when they come up against "establishment science."

Extrasensory Perception and Psychokinesis

Honorton, C. Relationship between EEG alpha activity and ESP card-guessing performance. *Journal of the American Society for Psychical Research*, 1969, 63, 365-374.

Ostrander, S., and Schroeder, L. *Psychic Discoveries Behind the Iron Curtain*. Englewood Cliffs, N.J.: Prentice-Hall, 1970.

Two touring reporters bring back some fascinating accounts of parapsychological research behind the iron curtain.

Targ, R., and Hurt, D. Learning clairvoyance and precognition with an extrasensory perception teaching machine. Paper presented at the Parapsychology Association Meeting, Durham, North Carolina, 1971.

Using feedback to teach extrasensory perception.

An Organic Feedback Machine

Backster, C. Evidence of a primary perception in plant life. *International Journal of Parapsychology*, Winter, 1968, 329-348.

The man responsible for research into plant "perception" describes one of his major experiments.

Lawrence, L. Electronics and the living plant. *Electronics World*, October, 1969, 25-28.

For those interested in trying their hand at plant communication, this article contains a description of the machinery involved.

Does Man Play Dice with the Universe?

Pearce, J. *The Crack in the Cosmic Egg*. New York: Julian Press, 1971.

A book challenging our established concepts of science and reality. Of particular interest is the chapter on fire walkers (pp. 99-109).

BIBLIOGRAPHY

Albino, R., and Burnand, G. Conditioning of the alpha rhythm in man. *Journal of Experimental Psychology,* 1964, 67, 539-544.

Alexander, L. Hypnotically induced hallucinations: Their diagnostic and therapeutic utilization. *Diseases of the Nervous System,* 1971, 32, 89-94.

Alger, I. Therapeutic use of videotape playback. *Journal of Nervous and Mental Disease,* 1969, 148, 430-436.

Alpha wave of the future. *Time,* July 19, 1971, 33.

Anand, B., Chhina, G., and Singh, B. Some aspects of electro-encephalographic studies in Yogis. *Electroencephalography and Clinical Neurophysiology,* 1961, 13, 452-456.

Anderson, S. Warm hands mean a cool, quiet head. *Midway* (the magazine of the *Topeka Capital-Journal*), August 15, 1971.

Andrews, L., and Karlins, M. *Requiem for Democracy?* New York: Holt, Rinehart & Winston, 1971.

Antrobus, J., and Antrobus, J. Discrimination of two sleep stages by human subjects. *Psychophysiology,* 1967, 4, 48-55.

————, ————, and Fisher, C. Discrimination of dreaming and nondreaming sleep. *Archives of General Psychiatry,* 1965, 12, 395-401.

Aserinsky, E., and Kleitman, N. Regularly occurring periods of eye motility and concomitant phenomena during sleep. *Science,* 1953, 118, 273-274.

————— and —————. Two types of ocular motility during sleep. *Journal of Applied Physiology,* 1955, 8, 1-10.

August, R. Hallucinatory experiences utilized for obstetric hypno-anesthesia. *American Journal of Clinical Hypnosis,* 1960, 3, 90-94.

Backster, C. Evidence of a primary perception in plant life. *International Journal of Parapsychology,* Winter, 1968, 329-348.

Bagchi, B., and Wenger, M. Electrophysiological correlates of some Yogi exercises. *Electroencephalography and Clinical Neurophysiology,* 1957 (Supplement 7), 132-149.

Bakan, P. The eyes have it. *Psychology Today,* April, 1971, 64+.

Barber, T., et al. (eds.). *Biofeedback and Self-Control, 1970.* Chicago: Aldine-Atherton, 1971.

————— et al. (eds.). *Biofeedback and Self-Control, 1971.* Chicago, Aldine-Atherton, 1972.

—————. Physiological effects of hypnosis and suggestion. In: T. Barber et al. (eds.). *Biofeedback and Self-Control, 1970.*

—————. Who believes in hypnosis? *Psychology Today,* July, 1970, 20+.

Barratt, P., and Herd, J. Subliminal conditioning of alpha rhythm. *Australian Journal of Psychology,* 1964, 16, 9-19.

Barrett, B. Reduction in rate of multiple tics by free operant conditioning methods. *Journal of Nervous and Mental Disease,* 1962, 135, 187-195.

Basmajian, J. Control and training of individual motor units. *Science,* 1963, 141, 440-441.

—————. *Muscles Alive, Their Functions Revealed by Electromyography.* Baltimore: Williams & Wilkins, 1967.

—————, Baeza, M., and Fabrigar, C. Conscious control and training of individual spinal motor neurons in normal human subjects. *Journal of New Drugs,* 1965, 5, 78-85.

————— and Simard, T. Effects of distracting movements on the control of trained motor units. *American Journal of Physical Medicine,* 1967, 46, 1427-1449.

Beal, J. Paraphysics and parapsychology: Recent developments associated with bio-electric field effects. Unpublished manuscript, 1971.

Beatty, J. Effects of initial alpha wave abundance and operant training procedures on occipital alpha and beta wave activity. *Psychonomic Science,* 1971, 23, 197-199.

Beckman, F., and Stein, M. A note on the relationship between per cent alpha time and efficiency in problem solving.
Journal of Psychology, 1961, 51, 169-172.

Beecher, H. Control of suffering in severe trauma. *Journal of the American Medical Association,* 1960, 173, 534-536.

Begleiter, H., and Platz, A. Evoked potentials: Modification by classical conditioning. *Science,* 1969, 166, 769-771.

Behavior mod for asthmatics. *Behavior Today,* September 27, 1971. 2.

Benson, H. Letter to the Chairman of the hearings before the Select Committee on Crime, House of Representatives, Serial No. 92-1 (Exhibit No. 31), June 2-4, 1971.

————, Herd, J., Morse, W., and Kelleher, R. Behavioral induction of arterial hypertension and its reversal. *American Journal of Physiology,* 1969, 217, 30-34.

————, Shapiro, D., Tursky, B., and Schwartz, G. Decreased systolic blood pressure through operant conditioning techniques in patients with essential hypertension. *Science,* 1971, 173, 740-742.

Bergman, J., and Johnson, H. The effects of instructional set and autonomic perception on cardiac control. *Psychophysiology,* 1971, 8, 180-190.

Bernal, M. Behavioral feedback in the modification of brat behaviors. *Journal of Nervous and Mental Disease,* 1969, 148, 375-385.

Birk, L., Crider, A., Shapiro, D., and Tursky, B. Operant electrodermal conditioning under partial curarization. *Journal of Comparative and Physiological Psychology,* 1966, 62, 165-166.

Black, A. The direct control of neural processes by reward and punishment. *American Scientist,* 1971, 59, 236-245.

————. The operant conditioning of central nervous system electrical activity. In: G. Bower (ed.). *The Psychology of Learning and Motivation* (Vol. 6), New York: Academic Press, 1972, in press.

164

Block, J., Lagerson, J., Zohman, L., and Kelly, G. A feed-back device for teaching diaphragmatic breathing. *American Review of Respiratory Disease*, 1969, 100, 577-578.

Bogen, J., and Gordon, H. Musical tests for functional later-alization with intracarotid amobarbital. *Nature*, 1971, 230, 524-525.

Booker, H., Rubow, R., and Coleman, P. Simplified feed-back in neuromuscular retraining: An automated ap-proach using electromyographic signals. *Archives of Physical Medicine and Rehabilitation*, 1969, 50, 621-625,

Brener, J., and Hothersall, D. Heart rate control under con-ditions of augmented sensory feedback. *Psychophysiol-ogy*, 1966, 3, 23-28.

———— and Kleinman, R. Learned control of decreases in systolic blood pressure. *Nature*, 1970, 226, 1063-1064.

————, ————, and Goesling, W. The effects of different exposures to augmented sensory feedback on the con-trol of heart rate. *Psychophysiology*, 1969, 5, 510-516.

Broadhurst, A., and Glass, A. Relationship of personality measures to the alpha rhythm of the electroencephalo-gram. *British Journal of Psychiatry*, 1969, 115, 199-204.

Brown, B. Awareness of EEG-subjective activity relation-ships detected within a closed feedback system. *Psycho-physiology*, 1971, 7, 451-464.

————. Recognition of aspects of consciousness through association with EEG alpha activity represented by a light signal. *Psychophysiology*, 1970, 6, 442-452.

Brown, C., and Wagman, A. Operant salivary conditioning in man. *Conditional Reflex*, 1968, 3, 128-129.

Budzynski, T., and Stoyva, J. Biofeedback techniques in behavior therapy and autogenic training. Unpublished manuscript, University of Colorado Medical Center, 1971.

———— and ————. An instrument for producing deep mus-cle relaxation by means of analog information feedback. *Journal of Applied Behavior Analysis*, 1969, 2, 231-237.

————, ————, and Adler, C. Feedback-induced muscle relaxation: Application to tension headaches. *Journal of Behavior Therapy and Experimental Psychiatry,* 1970, 1, 205-211.

Chapman, R., Cavonius, L., and Ernest, J. Alpha and kappa EEG activity in eyeless subjects. *Science,* 1971, 171, 1159-1160.

Childers, D., and Perry, N. Alpha-like activity in vision. *Brain Research,* 1971, 25, 1-20.

Christy, A., and Vitale, J. Operant conditioning of high blood pressure: A pilot study. Unpublished manuscript, San Francisco Veterans Administration Hospital, 1971.

Clemente, C. Comments on the brain as an effector organ for the study of conditional reflexes. *Conditional Reflex,* 1970, 5, 153-155.

Collier, B. Brain power: The case for bio-feedback training. *Saturday Review,* April 10, 1971, 10+.

Controlling the inner man. *Time,* July 18, 1969.

Costa, L., Cox, M., and Katzman, R. Relationship between MMPI variables and percentage and amplitude of EEG alpha activity. *Journal of Consulting Psychology,* 1965, 29, 90.

Crider, A., Schwartz, G., and Shapiro, D. Operant suppression of electrodermal response rate as a function of punishment schedule. *Journal of Experimental Psychology,* 1970, 83, 333-334.

————, ————, and Shnidman, S. On the criteria for instrumental autonomic conditioning: A reply to Katkin and Murray. *Psychological Bulletin,* 1969, 71, 455-461.

————, Shapiro, D., and Tursky, B. Reinforcement of spontaneous electrodermal activity. *Journal of Comparative and Physiological Psychology,* 1966, 61, 20-27.

Darrow, C., and Gullickson, G. The role of brain waves in learning and other integrative functions. *Recent Advances in Biological Psychiatry,* 1968, 10, 249-256.

Datey, K., Deshmukh, S., Dalvi, C., and Vinekar, S. "Shavasan": a yogic exercise in the management of hypertension. *Angiology,* 1969, 20, 325-333.

Dean, S., Martin, R., and Streiner, D. Mediational control of the GSR. *Journal of Experimental Research in Personality,* 1968, 3, 71-76.

Deikman, A. Experimental mediation. *Journal of Nervous and Mental Disease,* 1963, 136, 329-343.

————. Implications of experimentally induced contemplative meditation. *Journal of Nervous and Mental Disease*, 1966, 142, 101-116.

Dement, W., and Kleitman, N. The relation of eye movements during sleep to dream activity: An objective method for the study of dreaming. *Journal of Experimental Psychology*, 1957, 53, 339-346.

Dewey, J. *Freedom and Culture.* New York: Putnam, 1939. electroencephalogram. *Proceedings of the Symposium of Biomedical Engineering*, 1966, 1, 349-351.

Dewey, J. Freedom and Culture. New York: Putnam, 1939.

DiCara, L. Instrumental learning of visceral and glandular responses and implications for psychosomatic medicine. *Proceedings*, 76th Annual Convention, American Psychological Association, 1968, 259-260.

————. Learning in the autonomic nervous system. *Scientific American*, 1970, 222, 30-39.

————. Learning of cardiovascular responses: A review and a description of physiological and biochemical consequences *Transactions of the New York Academy of Sciences*, 1971, 33, 411-422.

———— and Miller, N. Instrumental learning of systolic blood pressure responses by curarized rats: Dissociation of cardiac and vascular changes. *Psychosomatic Medicine*, 1968, 30, 489-494.

———— and ————. Instrumental learning of vasomotor responses by rats: Learning to respond differently in the two ears. *Science*, 1968, 159, 1485-1486.

Dimond, S., and Beaumont, G. Use of two cerebral hemispheres to increase brain capacity. *Nature*, 1971, 232, 270-271.

Eisdorfer, C., Nowlin, J., and Wilkie, F. Improvement of learning in the aged by modification of autonomic nervous system activity. *Science*, 1970, 170, 1327-1329.

Eisendrath, R. The role of grief and fear in the death of kidney transplant patients. *American Journal of Psychiatry*, 1969, 126, 381-387.

Engel, B., and Chism, R. Operant conditioning of heart rate speeding. *Psychophysiology*, 1967, 3, 418-426.

———— and Hansen, S. Operant conditioning of heart rate slowing. *Psychophysiology*, 1966, 3, 176-187.

————, and Melmon, K. Operant conditioning of heart rate in patients with cardiac arrhythmias. *Conditional Reflex,* 1968, 3, 130.

Engstrom, D., London, P., and Hart, J. Hypnotic susceptibility increased by EEG alpha training. *Nature,* 1970, 227, 1261-1262.

Erickson, M. The interspersal hypnotic technique for symptom correction and pain control. *American Journal of Clinical Hypnosis,* 1966, 8, 198-209.

Evans, F. An experimental indirect technique for the induction of hypnosis without awareness. *International Journal of Clinical and Experimental Hypnosis,* 1967, 15, 72-85.

Fadiman, J. The Council Grove Conference on altered states of consciousness. *Journal of Humanistic Psychology,* 1969, 9, 135-137.

Feather, B. Human salivary conditioning. *Conditional Reflex,* 1968, 3, 129.

Fehmi, L. Bio-feedback of electroencephalographic parameters and related states of consciousness. Paper delivered at the American Psychological Association Meeting, Washington, D.C., September, 1971.

Fenton, G., and Scotton, L. Personality and the alpha rhythm. *British Journal of Psychiatry,* 1967, 113, 1283-1289.

Fernandez, H., Robinson, R., and Taylor, R. A device for testing consciousness. *American Journal of EEG Technology,* 1967, 7, 77-78.

Fetz, E. Operant conditioning of cortical unit activity. *Science,* 1969, 163, 955-958.

Filion, R., Fowler, S., and Notterman, J. Psychophysical evaluation of feedback phenomena as related to precision of force emission: Some methodological considerations. *American Journal of Psychology,* 1969, 82, 266-271.

Fisher, C., Gross, J., and Zuch, J. Cycle of penile erection synchronous with dreaming (REM) sleep. *Archives of General Psychiatry,* 1965, 12, 29-45.

Foulkes, D., and Vogel, G. Mental activity at sleep onset. *Journal of Abnormal Psychology,* 1965, 70, 231-243.

Fowler, R., and Kimmel, H. Operant conditioning of the GSR. *Journal of Experimental Psychology*, 1962, 63, 563-567.

Fox, S., and Rudell, A. Operant controlled neural event: Formal and systematic approach to electrical coding of behavior in brain. *Science*, 1968, 162, 1299-1302.

Fruhling, M., Basmajian, J., and Simard, T. A note on the conscious controls of motor units by children under six. *Journal of Motor Behavior*, 1969, 1, 65-68.

Galbraith, G., London, P., Leibovitz, M., Cooper, L., and Hart, J. EEG and hypnotic susceptibility. *Journal of Comparative and Physiological Psychology*, 1970, 72, 125-131.

Gavalas, R. Operant reinforcement of an autonomic response: Two studies. *Journal of the Experimental Analysis of Behavior*, 1967, 10, 119-130.

Giannitrapani, D. EEG average frequency and intelligence. *Electroencephalography and Clinical Neurophysiology*, 1969, 27, 480-486.

Glass, A. Intensity of attenuation of alpha activity by mental arithmetic in females and males. *Physiology and Behavior*, 1968, 3, 117-220.

Glassman, J. Alpha waves: The wave of the future. *Metropolitan Review*, October 5, 1971, 5+.

Gordon, T. Bucking the scientific establishment. *Playboy*, April, 1968, 127+.

Gould, D. All a matter of brain waves. *New Statesman*, January 17, 1969, 77.

Gray, E. Conscious control of motor units in a tonic muscle. *American Journal of Physical Medicine*, 1971, 52, 34-40.

Green, E., and Green, A. Conference on voluntary control of internal states. *Psychologia: An International Journal of Psychology in the Orient*, 1969, 12, 107-108.

——— and ———. On the meaning of transpersonal: Some metaphysical perspectives. *Journal of Transpersonal Psychology*, 1971, 3, 27-46.

———, ———, and Walters, E. Self-regulation of internal states. In: J. Rose (ed.). *Progress of Cybernetics: Proceedings of the International Congress of Cybernetics, London, 1969*. London: Gordon and Breach, 1970.

————, ————, and ————. Voluntary control of internal states: Psychological and physiological. *Journal of Transpersonal Psychology*, 1970, 2, 1-26.

————, Ferguson, D., Green, A., and Walters, E. Preliminary report on voluntary controls project: Swami Rama. Unpublished manuscript, the Menninger Foundation, 1970.

————, Walters, E., Green, A., and Murphy, G. Feedback technique for deep relaxation. *Psychophysiology*, 1969, 6, 371-377.

Grim, P. Anxiety change produced by self-induced muscle tension and by relaxation with respiration feedback. *Behavior Therapy*, 1971, 2, 11-17.

Hardyck, C., and Petrinovich, L. Subvocal speech and comprehension level as a function of the difficulty level of reading material. *Journal of Verbal Learning and Verbal Behavior*, 1970, 9, 647-652.

———— and ————. Treatment of subvocal speech during reading. *Journal of Reading*, February, 1969, 361-368+

————, ————, and Ellsworth, D. Feedback of speech muscle activity during silent reading: Rapid extinction. *Science*, 1966, 154, 1467-1468.

Hare, R., and Quinn, M. Psychopathy and autonomic conditioning. *Journal of Abnormal Psychology*, 1971, 77, 223-235.

Harrison, V., and Mortensen, O. Identification and voluntary control of single motor unit activity in the tibialis muscle. *Anatomical Record*, 1962, 144, 109-116.

Hart, J. Autocontrol of EEG alpha. *Psychophysiology*, 1968, 4, 506.

————. Beyond psychotherapy—a programmatic essay on the applied psychology of the future. In: T. Barber et al. (eds.). *Biofeedback and Self-Control, 1970.*

Headrick, M., Feather, B., and Wells, D. Unidirectional and large magnitude heart rate changes with augmented sensory feedback. *Psychophysiology*, 1971, 8, 132-142.

Henahan, D. Music draws strains direct from brains. *New York Times*, November 25, 1970.

Hilgard, E. Altered states of awareness. *Journal of Nervous and Mental Disease*, 1969, 149, 68-79.

Hinman, A., Engel, B., and Bickford, A. Portable blood

pressure recorder: Accuracy and preliminary use in evaluating intra-daily variations in pressure. *American Heart Journal*, 1962, 63, 663-668.

Hnatiow, M., and Lang, P. Learned stabilization of cardiac rate. *Psychophysiology*, 1965, 1, 330-336.

Hoenig, J. Medical research on yoga. *Confinia Psychiatrica*, 1968, 11, 69-89.

Honorton, C. Relationship between EEG alpha activity and ESP card-guessing performance. *Journal of the American Society for Psychical Research*, 1969, 63, 365-374.

Hord, D., and Barber, J. Alpha control: Effectiveness of two kinds of feedback. *Psychonomic Science*, in press.

How much for your alpha machine? *Behavior Today*, September 20, 1971.

Human medicine. *Behavior Today*, May 31, 1971.

Jacobs, A., and Felton, G. Visual feedback of myoelectric output to facilitate muscle relaxation in normal persons and patients with neck injuries. *Archives of Physical Medicine and Rehabilitation*, 1969, 50, 34-39.

Jacobson, E. *Progressive Relaxation*. Chicago: University of Chicago Press, 1938.

James, W. *The Varieties of Religious Experience*. New York: Mentor Books, 1958.

Jasper, M., and Shagass, C. Conditioning the occipital alpha rhythm in man. *Journal of Experimental Psychology*, 1941, 28, 373-388.

Johnson, H., and Schwartz, G. Suppression of GSR activity through operant reinforcement. *Journal of Experimental Psychology*, 1967, 75, 307-312.

Johnson, L., and Hord, D. Recuperative value of self-regulated states and of brief sleep for sleep loss and fatigue. Progress report, Navy Medical Neuropsychiatric Research Unit (San Diego, California), 1971.

Kalech, M. Now—a zen machine. *New York Post*, September 23, 1971.

Kamiya, J. Behavioral and physiological concomitants of dreaming. Progress report, National Institute of Health Grants M-2116 and M-5069, February, 1962.

———. Conscious control of brain waves. *Psychology Today*, 1968, 1, 56-60.

———. A fourth dimension of consciousness. *Journal of Experimental Medicine and Surgery*, 1969, 27, 13-18.

———. Operant control of the EEG alpha rhythm and some of its reported effects on consciousness. In: C. Tart (ed.). *Altered States of Consciousness*, 507-517.

Kasamatsu, A., and Hirai, T. An electroencephalographic study on the Zen meditation (Zaren). In: C. Tart (ed.). *Altered States of Consciousness*, 489-501.

Katkin, E., and Murray, E. Instrumental conditioning of autonomically mediated behavior: Theoretical and methodological issues. *Psychological Bulletin*, 1968, 70, 52-68.

———, ———, and Lachman, R. Concerning instrumental autonomic conditioning: A rejoinder. *Psychological Bulletin*, 1969, 71, 462-466.

Kennedy, J. A possible artifact in electroencephalography. *Psychological Review*, 1959, 66, 347-352.

Kiefer, D. Meditation and biofeedback. In: J. White (ed.). *The Highest State of Consciousness*.

Kimberly, R. Rhythmic patterns in human interaction. *Nature*, 1970, 228, 88-90.

Kimmel, E., and Kimmel, H. A replication of operant conditioning of the GSR. *Journal of Experimental Psychology*, 1963, 65, 212-213.

Kimmel, H. Instrumental conditioning of autonomically mediated behavior. *Psychological Bulletin*, 1967, 67, 337-345.

Koegler, R., Hicks, S., and Barger, J. Medical and psychiatric use of electrosleep. *Diseases of the Nervous System*, 1971, 32, 100-104.

Korein, J., Maccario, M., Carmona, A., Randt, C., and Miller, N. Operant conditioning techniques in normal and abnormal EEG states. Paper presented at the American Academy of Neurology Meeting, April, 1971.

Lacey, J., and Lacey, B. The law of initial value in the longitudinal study of autonomic constitution: Reproducibility of autonomic responses and response patterns over a four-year interval. *Annals New York Academy of Sciences*, 1962, 98, 1257-1290.

Lang, P. Autonomic control or learning to play the internal organs. *Psychology Today*, October, 1970, 37-41+.

——— and Melamed, B. Case report: Avoidance condition-iting. *Journal of Abnormal Psychology*, 1969, 74, 1-8.
iting. *Journal of Abnormal Psychology*, 1969, 74, 1-8.
———, Sroufe, A., and Hastings, J. Effects of feedback and instructional set on the control of cardiac-rate variability. *Journal of Experimental Psychology*, 1967, 75, 425-431.

Lawrence, L. Electronics and the living plant. *Electronics World*, October, 1969, 25-28.

Laws, D., and Rubin, H. Instructional control of an auto-nomic sexual response. *Journal of Applied Behavior Analysis*, 1969, 2, 93-99.

Lesh, T. Zen meditation and the development of empathy in counselors. In: T. Barber et al (eds.). *Biofeedback and Self-Control, 1970.*

Levenet, H., Engel, B., and Pearson, J. Differential operant conditioning of heart rate. *Psychosomatic Medicine*, 1968, 30, 837-845.

Levy, J. Possible basis for the evolution of laternal speciali-zation of the human brain. *Nature*, 1969, 224, 614-615.

Lewinsohn, P., and Shaw, D. Feedback about interpersonal behavior as an agent of behavior change. *Psychotherapy and Psychosomatics*, 1969, 17, 82-88.

Lippold, O., and Redfearn, W. Mental changes resulting from the passage of small direct currents through the human brain. *British Journal of Psychiatry*, 1964, 110, 768-772.

——— and Shaw, J. Alpha rhythm in the blind. *Nature*, 1971, 232, 134.

Loehr, F. *The Power of Prayer on Plants.* New York: New American Library (Signet Books), 1969.

London, P. *Behavior Control.* New York: Harper & Row, 1969.

———, Hart, J., and Leibovitz, M. EEG alpha rhythms and susceptibility to hypnosis. *Nature*, 1968, 219, 71-72.

Lowinger, P., and Dobie, S. What makes the placebo work? *Archives of General Psychiatry*, 1969, 20, 84-88.

Lubin, A., Johnson, L., and Austin, M. Discrimination among states of consciousness using EEG spectra. *Psychophysi-ology*, 1969, 6, 122-132.

Luce, G. *Body Time.* New York: Pantheon, 1971.

———. Understanding body time in the 24-hour city. *New York Magazine*, November 15, 1971, 37+.

——— and Peper, E. Biofeedback: Mind over body, mind

173

over mind. *New York Times Magazine*, September 12, 1971, 34+.

—— and Segal, J. *Sleep*. New York: Coward-McCann, 1966.

Lynch, J., Orne, M., Paskewitz, D., and Costello, J. An analysis of the feedback control of alpha activity. *Contitional Reflex*, 1970, 5, 185-186.

—— and Paskewitz, D. On the mechanism of the feedback control of human brain wave activity. *Journal of Nervous and Mental Disease*, 1971, 153, 205-217.

McCain, G., and Segal, E. *The Game of Science*. Belmont, Calif.: Brooks-Cole, 1969.

McFall, R. Effects of self-monitoring on normal smoking behavior. *Journal of Consulting and Clinical Psychology*, 1970, 35, 135-142.

McGraw, W. Plants are only human. *Argosy*, June, 1969, 24-27.

McGuigan, F., Camacho, E., Hardyck, C., Petrinovich, L., and Ellsworth, D. Feedback of speech muscle activity during silent reading: Two comments. *Science*, 1967, 157, 579-581.

Malmo, R. Emotions and muscle tension: The story of Anne. *Psychology Today*, March, 1970, 64-67+.

Man into superman: The promise and peril of the new genetics. *Time*, April 19, 1971.

Marsden, C., Meadows, J., and Hodgson, H. Observations on the reflex response to muscle vibration in man and its voluntary control. *Brain*, 1969, 92, 829-846.

Mason, J. Strategy in psychosomatic research. In: T. Barber et al. (eds.). *Biofeedback and Self-Control*.

Mathews, A., and Gelder, M. Psycho-physiological investigations of brief relaxation training. *Journal of Psychosomatic Research*, 1969, 13, 1-12.

May, J., and Johnson, H. Positive reinforcement and suppression of spontaneous GSR activity. *Journal of Experimental psychology*, 1969, 80, 193-195.

Mayr, O. The origins of feedback control. *Scientific American*, 1970, 223, 110-118.

Miller, H. Alpha waves—artifacts? *Psychological Bulletin*, 1968, 69, 279-280.

Miller, N. Learning of visceral and glandular responses. *Science,* 1969, 163, 434-445.

———. Psychosomatic effects of specific types of training. *Annals of the New York Academy of Sciences,* 1969, 159, 1025-1040.

——— and Banuazizi, A. Instrumental learning by curarized rats of a specific visceral response, intestinal or cardiac. *Journal of Comparative and Physiological Psychology,* 1968, 65, 1-7.

——— and DiCara, L. Instrumental learning of heart rate changes in curarized rats: Shaping, and specificity to discriminative stimulus. *Journal of Comparative and Physiological Psychology,* 1967, 63, 12-19.

——— and ———. Instrumental learning of urine formation by rats; changes in renal blood flow. *American Journal of Physiology,* 1968, 215, 677-683.

———, ———, and Banuazizi, A. Instrumental learning of glandular and visceral responses. *Conditional Reflex,* 1968, 3, 129.

———, ———, Solomon, H., Weiss, J., and Dworkin, M. Learned modifications of autonomic functions: A review and some new data. *Circulation Research* (Supplement 1), 1970, 26-27, I-3–I-11.

Mind and ulcer. *British Medical Journal,* August 16, 1969, 374.

Mind over drugs. *Time,* October 25, 1971, 51.

Moonstruck scientists. *Time,* January 10, 1972.

Morgenson, D., and Martin, I. Personality, awareness, and autonomic conditioning. *Psychophysiology,* 1969, 5, 536-547.

Mulholland, T. Can you really turn on with alpha? Paper presented at the meeting of the Massachusetts Psychological Association, May 7, 1971.

———. The concept of attention and the electroencephalographic alpha rhythm. In: C. Evans and T. Mulholland (eds.). *Attention in Neurophysiology.* London: Butterworths, 1969.

———. Feedback electroencephalography. *Activitas Nervosa Superior* (Prague), 1968, 10, 410-438.

———. Occipital alpha revisited. In press: *Psychological Bulletin,* 1972.

——— and Benson, D. Feedback electroencephalography in

clinical research. Unpublished manuscript, Bedford, Massachusetts, Veterans Administration Hospital, 1971.

————— and Peper, E. Occipital alpha and accommodative vergence, pursuit tracking, and fast eye movements. *Psychophysiology*, 1971, 8, 556, 575.

————— and Runnals, S. Evaluation of attention and alertness with a stimulus-brain feedback loop. *Electroencephalography and Clinical Neurophysiology*, 1962, 14, 847-852.

Murphy, G. Experiments in overcoming self-deception. In: T. Barber et al. (eds.). *Biofeedback and Self-Control, 1970.*

—————. Psychology in the year 2000. *American Psychologist*, 1969, 24, 523-530.

Muscle is retrained at home. *Medical World News*, December 10, 1971, 35.

Naranjo, C., and Ornstein, R. *On the Psychology of Meditation.* New York: Viking, 1971.

Needleman, J. *The New Religions.* New York: Doubleday, 1970.

Notterman, J., and Fowler, S. Toward a broader description of operant behavior. Summary of research funded under National Institute of Mental Health Grant 18189, December, 1970.

Nowlis, D., and Kamiya, J. The control of electroencephalographic alpha rhythms through auditory feedback and the associated mental activity. *Psychophysiology*, 1970, 6, 476-484.

————— and Rhead, J. Relation of eyes-closed resting EEG alpha activity to hypnotic susceptibility. *Perceptual and Motor Skills*, 1968, 27, 1047-1050.

Obrist, P., Webb, R., Sutterer, J., and Howard, J. The cardiacsomatic relationship: Some reformulations. *Psychophysiology*, 1970, 6, 569-587.

Okeima, T., Kogu, E., Ideda, K., and Sugiyama, H. The EEG of Yoga and Zen practitioners. *Electroencephalography and Clinical Neurophysiology*, 1957, 51 (Supplement 9).

Olds, J. The central nervous system and the reinforcement of behavior. *American Psychologist*, 1969, 24, 114-132.

O'Leary, J. Discoverer of the brain wave. *Science*, 1970, 168, 562-563.

Ostrander, S., and Schroeder, L. *Psychic Discoveries Behind the Iron Curtain.* Englewood Cliffs, N.J.: Prentice-Hall, 1970.

Paredes, A., Ludwig, K., Hassenfeld, I., and Cornelison, F. A clinical study of alcoholics using audiovisual self-image feedback. *Journal of Nervous and Mental Disease,* 1969, 148, 449-456.

Paskewitz, D., Lynch, J., Orne, M., and Costello, J. The feedback control of alpha activity: Conditioning or disinhibition? *Psychophysiology,* 1970, 6, 637-638.

Pasquali, E. A relay controlled by alpha rhythm. *Psychophysiology,* 1969, 6, 207-208.

Passerini, D., and Paterson, S. A study of cardiac conditioning in man. *Conditional Reflex,* 1966, 1, 90-103.

Pearce, J. *The Crack in the Cosmic Egg.* New York: Julian Press, 1971.

Pearlman, C., and Greenberg, R. Medical-psychological implications of recent sleep research. *Psychiatry in Medicine,* 1970, 1, 261-276.

Peper, E. Comment on feedback training of parietal-occipital alpha asymmetry in normal human subjects. *Kybernetik,* 1971, 9, 156-158.

———. Developing a biofeedback model: Alpha EEG feedback as a means for pain control. Unpublished manuscript, Bedford, Massachusetts, Veterans Administration Hospital, 1972.

———. Feedback regulation of the alpha electroencephalogram activity through control of the internal and external parameters. *Kybernetik,* 1970, 7, 107-112.

———. Reduction of efferent motor commands during alpha feedback as a precondition for changes in consciousness. *Kybernetik,* 1972, in press.

——— and Mulholland, T. Methodological and theoretical problems in the voluntary control of electroencephalographic occipital alpha by the subject. *Kybernetik,* 1970, 7, 10-13.

Phillips, D. Dying as a form of social behavior. Doctoral dissertation, Princeton University, 1969.

Pillsbury, J., Meyerowitz, S., Salzman, L., and Satran, R. Electroencephalographic correlates of perceptual style: Field Orientation. *Psychosomatic Medicine,* 1967, 29, 441-449.

Plumlee, L. Operant conditioning of increases in blood pressure. *Psychophysiology*, 1969, 6, 283-290.

Probing the brain. *Newsweek*, June 21, 1971, 60-67.

Quinn, J., Harbison, J., and McAllister, A. An attempt to shape human penile responses. *Behavior Research and Therapy*, 1970, 8, 213-216.

Ramsay, J., and Schlagenhauf, G. Treatment of depression with low voltage direct current. *Southern Medical Journal*, 1966, 59, 932-934.

Randle, R. Volitional control of visual accommodation. Paper presented at ASMP, AGARD, NATO, Garmisch-Partenkirchen, Germany, September, 1970.

Razran, G. The observable unconscious and the inferable conscious in current Soviet psychophysiology: Interoceptive conditioning, semantic conditioning and the orienting reflex. *Psychological Review*, 1961, 68, 81-147.

Rogers, J. Drug abuse—just what the doctor ordered. *Psychology Today*, September, 1971, 16+.

Rorvik, D. The wave of the future: Brain waves. *Look*, October 6, 1970, 88+.

Rosenberg, B. Turning on—electroencephalographically. *Smith, Kline and French Psychiatric Reporter*, 1969, 45, 9-11.

Rosenfeld, J., Rudell, A., and Fox, S. Operant control of neural events in humans. *Science*, 1969, 165, 821-823.

Rosenthal, S., and Wulfsohn, N. Electrosleep. A preliminary communication. *Journal of Nervous and Mental Disease*, 1970, 151, 146-151.

Rubow, R., and Smith, K. Feedback parameters of electromyographic learning. *American Journal of Physical Medicine*, 1971, 50, 115-131.

Russ, K., Santorum, A., and Lanyon, R. Operant conditioning of vasodilation in patients with essential hypertension. Unpublished manuscript, Butler, Pennsylvania, Veterans Administration Hospital, 1971.

Sanford, E. An acoustic mirror in psychotherapy: The audiotape recorder as a psychotherapeutic tool. *American Journal of Psychotherapy*, 1969, 23, 681-695.

Sargent, J., Green, E., and Walters, E. Preliminary report on the use of autogenic feedback techniques in the treatment of migraine and tension headaches. Unpublished manuscript, Menninger Foundation, 1971.

Schafer, W. Further developments of the field effect monitor. Life Sciences, Corvair Division of General Dynamics, Report GDC-ERR-AN-1114, October, 1967.

Schmeidler, G. High ESP scores after a swami's brief instruction in meditation and breathing. *Journal of the American Society for Psychical Research,* 1970, 64, 100-103.

———. Respice, adspice, prospice. Presidential address to the Parapsychological Association, 1971.

——— and Lewis, L. Mood changes after alpha feedback training. *Perceptual and Motor Skills,* 1971, 32, 709-710.

Schultz, J., and Luthe, W. *Autogenic Training: A Physiologic Approach in Psychotherapy.* New York: Grune and Stratton, 1959.

Schwartz, G., and Johnson, H. Affective visual stimuli as operant reinforcers of the GSR. *Journal of Experimental Psychology,* 1969, 80, 28-32.

———, Shapiro, D., and Tursky, B. Learned control of cardiovascular integration in man through operant conditioning. *Psychosomatic Medicine,* 1971, 33, 57-62.

Schwitzgebel, R. Behavior instrumentation and social technology. *American Psychologist,* 1970, 25, 491-499.

———. Survey of electromechanical devices for behavior modification. *Psychological Bulletin,* 1968, 70, 444-459.

Scully, H., and Basmajian, J. Motor-unit training and influence of manual skill. *Psychophysiology,* 1969, 5, 625-632.

Shaffer, R. Bio-Feedback. *Wall Street Journal,* April 19, 1971.

Shapiro, D., and Crider, A. Operant electrodermal conditioning under multiple schedules of reinforcement. *Psychophysiology,* 1967, 4, 168-175.

———, ———, and Tursky, B. Classical conditioning and incubation of human diastolic blood pressure. *Conditional Reflex,* 1968, 3, 132-133.

——— and Schwartz, G. Psychophysiological contributions to social psychology. *Annual Review of Psychology,* 1970, 21, 87-112.

———, ———, and Tursky, B. Control of diastolic blood pressure in man by feedback and reinforcement. *Psychophysiology,* in press.

———, Tursky, B., Gershon, E., and Stern, M. Effects of

feedback and reinforcement on the control of human systolic blood pressure. *Science,* 1969, 163, 588-590.

————, ————, and Schwartz, G. Differentiation of heart rate and systolic blood pressure in man by operant conditioning. *Psychosomatic Medicine,* 1970, 32, 417-423.

Shearn, D. Operant conditioning of heart rate. *Science,* 1962, 137, 530-531.

Shipman, W., Oken, D., and Heath, H. Muscle tension and effort at self-control during anxiety. *Archives of General Psychiatry,* 1970, 23, 359-368.

Simpson, H., Paivio, A., and Rogers, T. Occipital alpha activity of high and low visual imagers during problem solving. *Psychonomic Science,* 1967, 8, 49-50.

Skinner, B. *Beyond Freedom and Dignity.* New York: Knopf, 1971.

————, *Walden Two.* New York: Macmillan, 1948.

Slater, K. Alpha rhythms and mental imagery. *Electroencephalography and Clinical Neurophysiology,* 1960, 12, 851-859.

Slucki, H., Adam, G., and Porter, R. Operant discrimination of an interoceptive stimulus in rhesus monkeys. *Journal of the Experimental Analysis of Behavior,* 1965, 8, 405-414.

————, McCoy, F., and Porter, R. Interoceptive S^D of the large intestine established by mechanical stimulation. *Psychological Reports,* 1969, 24, 35-42.

Smart, A. Conscious control of physical and mental states. *Menninger Perspective,* April-May, 1970.

Smith, K. Cybernetic foundations of behavioral science. (A collection of published articles.) University of Wisconsin Behavioral Cybernetics Laboratory, undated.

———— and Smith, T. Systems theory of therapeutic and rehabilitative learning with television (feedback). *Journal of Nervous and Mental Disease,* 1969, 148, 386-429.

Snyder, C., and Noble, M. Operant conditioning of vasoconstriction. *Journal of Experimental Psychology,* 1968, 77, 263-268.

Sokolow, M., Werdegar, D., Perloff, D., Cowan, R., and Brenenstuhl, H. Preliminary studies relating portably recorded blood pressures to daily life events in patients with essential hypertension. In: M. Koster et al (eds.).

Psychosomatics in Essential Hypertension. Basel (Switzerland): S. Karger, 1970.

Solomon, G. Emotions, stress and immunity. Unpublished manuscript, Stanford University School of Medicine, undated.

Sperry, R. A modified concept of consciousness. *Psychological Review,* 1969, 76, 532-536.

————. An objective approach to subjective experience: Further explanation of a hypothesis. *Psychological Review,* 1970, 77, 585-590.

Spilker, B., Kamiya, J., Callaway, E., and Yeager, C. Visual evoked responses in subjects trained to control alpha rhythms. *Psychophysiology,* 1969, 5, 683-695.

Stanford, R., and Lovin, C. The EEG alpha rhythm and ESP performance. *Journal of the American Society for Psychical Research,* 1970, 64.

Sterman, M. Effects of instrumental EEG conditioning upon sleep and seizure behavior in the cat. *Conditional Reflex,* 1970, 5, 185.

————, Howe, R., and MacDonald, L. Facilitation of spindle-burst sleep by conditioning of electroencephalographic activity while awake. *Science,* 1970, 167, 1146-1148.

Stern, R., and Lewis, N. Ability of actors to control their GSRs and express emotions. *Psychophysiology,* 1968, 4, 294-299.

Stoller, F. Group psychotherapy on television: An innovation with hospitalized patients. *American Psychologist,* 1967, 22, 158-162.

————. The long weekend. *Psychology Today,* December, 1967, 28-33.

Stoyva, J. The public (scientific) study of private events. In: T. Barber et al. (eds.). *Biofeedback and Self-Control, 1970.*

———— and Kamiya, J. Electrophysiological studies of dreaming as the prototype of a new strategy in the study of consciousness. *Psychological Review,* 1968, 75, 192-205.

Streiner, D., and Watters, D. Instrumental conditioning of the GSR in normals and schizophrenics. *Proceedings,* 79th Annual Convention, American Psychological Association, 1971.

Tani, K., and Yoshii, N. Efficiency of verbal learning during sleep as related to the EEG pattern. *Brain Research,* 1970, 17, 227-285.

Targ, R., and Hurt, D. Learning clairvoyance and precognition with an extrasensory perception teaching machine. Paper presented at the Parapsychology Association Meeting, Durham, North Carolina, 1971.

Tart, C. (ed.). *Altered States of Consciousness.* New York: Wiley, 1969.

————. A psychophysiological study of out-of-the-body experiences in a selected subject. *Journal of the American Society for Psychical Research,* 1968, 62, 3-27.

Thomson, D., and Ward, J. Human communication using fields generated by the nervous system. *Schizophrenia,* 1970, 2, 140-143.

Tiller, W. The A.R.E. visit to Russia. Unpublished manuscript, undated.

Turning on with alpha waves. *Life,* August 21, 1970, 60-61.

Ullman, M., and Krippner, S. ESP in the night. *Psychology Today,* June, 1970, 47+.

Vogel, W., Broverman, D., and Klaiber, E. EEG and mental abilities. *Electroencephalography and Clinical Neurophysiology,* 1968, 24, 166-175.

Wallace, R. Physiological effects of transcendental meditation. *Science,* 1970, 167, 1751-1754.

————, Benson, H., and Wilson, A. A wakeful hypometabolic physiologic state. *American Journal of Physiology,* 1971, 221, 795-799.

Walter, D., Rhodes, J., and Adey, W. Discriminating among states of consciousness by EEG measurements. A study of four subjects. *Electroencephalography and Clinical Neurophysiology,* 1967, 22, 2-29.

Walter, W. The social organ. *Impact of Science on Society,* 1968, 18, 179-186.

Washington, B. Child's "bellyache" was a family affair. *New York Post,* Monday, November 1, 1971.

Watanabi, T., Shapiro, D., and Schwartz, G. Meditation as an anoxic state: A critical review and theory. Paper presented at the Psychophysiology Society Meetings, St. Louis, Missouri, 1971.

Watson, J. *Behaviorism.* New York: People's Institute, 1924.

182

Weinberg, H. Correlation of frequency spectra of averaged visual evoked potentials with verbal intelligence. *Nature,* 1969, 224, 813-815.

Weiner, H. Current status and future prospects for research in psychosomatic medicine. *Journal of Psychiatric Research,* 1971, 8, 479-498.

Weiss, T., and Engel, B. Operant conditioning of heart rate in patients with premature ventricular contractions. *Psychosomatic Medicine,* 1971, 33, 301-321.

Wenger, M., and Bagchi, B. Studies of autonomic functions in practitioners of Yoga in India. *Behavioral Science,* 1961, 6, 312-323.

———, ———, and Anand, B. Experiments in India on "voluntary" control of the heart and pulse. *Circulation,* 1961, 24, 1319-1325.

Whatmore, G., and Kohli, D. Dysponesis: A neurophysiologic factor in functional disorders. *Behavioral Science,* 1968, 13, 102-124.

White, J. (ed.). *The Highest State of Consciousness.* New York: Doubleday (Anchor Original), 1972.

Wickramasekera, I. Effects of EMG feedback training on susceptibility to hypnosis: Preliminary observations. *Proceedings,* 79th Annual American Psychological Association Convention, 1971.

Williams, H. The new biology of sleep. *Journal of Psychiatric Research,* 1971, 8, 445-478.

Wilson, A., and Wilson, A. Psychophysiological and learning correlates of anxiety and induced muscle relaxation *Psychophysiology,* 1970, 6, 740-748.

Wilson, R. Cardiac response: Determinants of conditioning. *Journal of Comparative and Physiological Psychology* (monograph), 1969, 68, pt. 2, 1-23.

Witter, C. Drugging and schooling. *Transaction,* July-August, 1971, 30-34.

Wyrwicka, W., and Sterman, M. Instrumental conditioning of sensorimotor cortex EEG spindles in the waking cat. *Physiology and Behavior,* 1968, 3, 703-707.

INDEX

184

186

THE BEST OF THE BESTSELLERS
FROM WARNER BOOKS!

THE WOMAN'S DRESS FOR SUCCESS BOOK   (87-672, $3.95)
by John T. Molloy
The runaway bestseller by America's foremost clothing engineer which tells women who want to get ahead how to dress like a winner. "John Molloy will help put women in the boss's chair by sharing his advice on how to dress for a successful business career."
—Chicago Tribune

SINGLE by Harriet Frank, Jr.   (81-543, $2.50)
A brilliant, moving novel about the lives, loves, tragedies and dreams of four "ordinary" women searching for happiness, finding it, losing it, crying or rejoicing over it, starting over, hanging on, making do . . . and surviving.

ANNA HASTINGS by Allen Drury   (81-603, $2.50)
With the speed of a wire service teletype, Anna Hastings shot out of the press gallery to become the founder of Washington's leading newspaper. But she paid a lifelong price for her legendary success.

SPARE PARTS by David A. Kaufelt   (81-889, $2.50)
A young reporter suddenly inherits controlling interest in a world-famous hospital. The hospital's uncanny success with transplant operations spurs the new owner's curiosity until he discovers a macabre secret entwined in a network of madness and treachery. A bizarre thriller more shocking than "Coma."

THE BONDMASTER BREED by Richard Tresillian   (81-890, $2.50)
The dazzling conclusion to the epic of Roxborough plantation, where slaves are the prime crop and the harvest is passion and rage.

THE MINNESOTA CONNECTION   (90-024, $1.95)
by Al Palmquist with John Stone
The terrifying true story of teenage prostitution in the vicious pipeline between Minneapolis and New York, and of a tough preacher-cop's determination to break it.

W A Warner Communications Company

THE BEST OF THE BESTSELLERS
FROM WARNER BOOKS!

THE OTHER SIDE OF THE MOUNTAIN   (82-935, $2.25)
by E.G. Valens
Olympic hopeful Jill Kinmont faced the last qualifying race before
the 1956 Games—and skied down the mountain to disaster, never
to walk again. Now she had another kind of mountain to climb—
to become another kind of champion.

THE OTHER SIDE OF THE MOUNTAIN:
PART 2 by E.G. Valens   (82-463, $2.25)
Part 2 of the inspirational story of a young Olympic contender's
courageous climb from paralysis and total helplessness to a useful
life and meaningful marriage. An NBC-TV movie and serialized in
Family Circle magazine.

SYBIL by Flora Rheta Schreiber   (82-492, $2.25)
Over 5 million copies in print! A television movie starring Joanne
Woodward, Sally Field and Martine Bartlett! A true story more
gripping than any novel of a woman possessed by sixteen separate
personalities. Her eventual integration into one whole person makes
this a "fascinating book."—**Chicago Tribune**

A STRANGER IN THE MIRROR   (81-940, $2.50)
by Sidney Sheldon
This is the story of Toby Temple, superstar and super bastard,
adored by his vast TV and movie public, but isolated from real
human contact by his own suspicion and distrust. It is also the
story of Jill Castle, who came to Hollywood to be a star and discovered she had to buy her way with her body. When these two
married, their love was so strong it was—terrifying!

 A Warner Communications Company

Please send me the books I have selected.

Enclose check or money order only, no cash please. Plus 50¢ per
order and 10¢ per copy to cover postage and handling. N.Y. State
and California residents add applicable sales tax.

Please allow 4 weeks for delivery.

WARNER BOOKS
P.O. Box 690
New York, N.Y. 10019

Name ....

Address ....

City .... State .... Zip ....

_____ Please send me your free mail order catalog